The Game of Conservation

Ohio University Press Series in Ecology and History

James L. A. Webb, Jr., Series Editor

The Game of Conservation

International Treaties to Protect the World's Migratory Animals

Mark Cioc

OHIO UNIVERSITY PRESS
ATHENS

Ohio University Press, Athens, Ohio 45701
www.ohioswallow.com
© 2009 by Ohio University Press
All rights reserved

Printed in the United States of America
Ohio University Press books are printed on acid-free paper ⊗ ™

16 15 14 13 12 11 10 09   5 4 3 2 1

Library of Congress Cataloging-in-Publication Data
The game of conservation : international treaties to protect the world's migratory animals /
Mark Cioc.
    p. cm. — (Ohio University Press series in ecology and history)
Includes bibliographical references and index.
ISBN 978-0-8214-1866-6 (cloth : alk. paper) — ISBN 978-0-8214-1867-3 (pbk. : alk. paper)
1. Migratory animals—Conservation—International cooperation. 2. Wildlife conservation (In-
ternational law) 3. Animal welfare—Law and legislation. 4. Treaties.
QH75.G36 2009
591.56'8—dc22

2009037629

In memory of
Dale McNeely (1951–1978)
and
Richard Crago (1952–2006)

Great friendships never die.

Contents

Illustrations

Acknowledgments

I thank the National Archives and Records Administration in College Park, Maryland, and the National Archives at Kew Gardens, for assisting me in my research. The archival materials that I utilized for this book were often difficult to identify and locate, and I greatly appreciate the extra effort that their staffs made to assist me.

I especially thank Bruce Thompson, my friend and colleague at the University of California, Santa Cruz, for reading every draft of every chapter. He left his imprint on every page. I also thank the three anonymous reviewers who read the original manuscript. They provided a wealth of constructive criticisms and valuable suggestions that helped me enormously as I prepared my final draft. Thanks also go to Helen Cole and Yvette Delgado for their excellent work in preparing the graphs and illustrations. Finally, I extend a heartfelt thanks to the series editor, James Webb, and to Ohio University Press's Gillian Berchowitz and Rick Huard for guiding the manuscript through the publication process. This book is much stronger because of their engagement and insights.

Introduction

The way to hunt is for as long as you live against
as long as there is such and such an animal.

—Ernest Hemingway, *Green Hills of Africa*

THIS IS a short book with a straightforward premise. I argue that the major animal-protection treaties of the early twentieth century are best understood as international *hunting* treaties rather than as *conservation* treaties. By and large, prominent hunters and ex-hunters ("penitent butchers," in the words of their critics) were the guiding force behind the treaties, and these hunters were often far more concerned with the protection of specific hunting grounds and prized prey than with the safeguarding of entire habitats, ecosystems, or bioregions. Over time, wildlife managers and conservationists tried to tweak these treaties into full-fledged animal-protection agreements. They discovered, however, that the hunting ethos embedded in the treaty texts hampered their efforts, and after 1946, they began to push for new approaches based on new premises. The strengths and weaknesses of these early treaties and the impact they had (often inadvertently) on subsequent animal-protection accords comprise the main subject matter of this book.

Wildlife conservationists owe a large debt to Aldo Leopold's pioneering *Game Management* (1933), for it was this book more than any other that first articulated the parallels between sustainable agriculture and sustainable hunting. "Game management," he wrote, "is the art of making land produce sustained annual crops of wild game for recreational purposes." A professional forester, an avid hunter, and an innovative ecologist, Leopold took a practical approach to wildlife conservation in the United States: game animals should be cultivated, like wheat and corn, their numbers augmented for human consumption. "There are still those who shy at this prospect of a man-made game crop as at something artificial and therefore repugnant," he noted. "This attitude shows great taste but poor insight. Every head of wild life still alive in this country is already artificialized, in that its existence is conditioned by economic forces."[1]

Farmers had long ago developed a variety of techniques—seeding, weeding, irrigating, fertilizing, fallowing, and the like—to maximize their annual yields. "Game cropping," by contrast, was still in its infancy and the tools of the trade largely still experimental and in flux. "History shows that game management nearly always has its beginnings in the control of the hunting factor," Leopold observed in the staccato-like prose for which he was famous: "Other controls are added later. The sequence seems to be about as follows: 1. Restriction of hunting. 2. Predator control. 3. Reservation of game lands (as parks, forests, refuges, etc.). 4. Artificial replenishment (restocking and game farming). 5. Environmental controls (control of food, cover, special factors, and disease)."[2]

More than seventy years after it first appeared, *Game Management* is still widely read by wardens and foresters—and with good reason, for it is both a practical guide for preserving game animals and an early history of wildlife administration. Like many game managers before and since, however, Leopold largely overlooked one of the key tools of animal conservation: *international treaties.* Few game species reside solely within the borders of a single country. Most are mobile creatures that crisscross national frontiers according to their needs, living at certain times of the year in colder and more temperate regions and other times in warmer and equatorial ones. Hunting laws, predator control, forest reserves, game farming, and habitat manipulation are all indispensable tools of conservation, but they often have little practical value if neighboring countries do not take similar measures. Effective game management depends on interregional links, transnational cooperation, and international agreements.

Governments worldwide have signed nearly fifteen hundred environmental treaties and agreements over the past century, fully half of which address the question of wildlife protection directly or indirectly. Many are simple bilateral fishing agreements designed to protect a shared river or a common delta. Others entail complex multinational initiatives that attempt to protect individual species or animal groups across many contiguous and noncontiguous countries. Still others handle habitat protection across thousands of miles, sometimes affecting regions far removed from human settlements. Big or small, comprehensive or limited, bilateral or multilateral, each treaty testifies to the importance of transnational cooperation in the effort to protect the world's wildlife. Animals recognize no political borders: they feed and breed wherever they find suitable niches, and they move about the earth according to the dictates of habitat and climate, not human whim.[3]

This book analyzes several key animal-protection treaties that were signed in the first half of the twentieth century. Each was designed to protect one or more of the world's most commercially valuable migratory species. Each exerted a powerful influence on other treaties in other parts of the world. And each had a lasting impact on nature protection worldwide. First, I analyze the two treaties that led to the creation of Africa's nature parks: the Convention for the Preservation of Wild Animals, Birds, and Fish in Africa (1900) and the Convention Relative to the Preservation of Flora and Fauna in Their Natural State (1933). Then, I examine the two treaties that brought a halt to the slaughter of game birds in North America: the Convention for the Protection of Migratory Birds (1916), signed between the United States and Canada, and the Convention between the United States of America and the United States of Mexico for the Protection of Migratory Birds and Game Mammals (1936). Finally, I look at the three failed attempts to protect the world's whale stocks: the Convention for the Regulation of Whaling (1931), the International Agreement for the Regulation of Whaling (1937), and the International Convention for the Regulation of Whaling (1946). That many of the world's species today enjoy a modicum of protection from overhunting (or "overharvesting," as Leopold would prefer) is largely due to the collective impact of these treaties. That many species still hover on the brink of extinction is also largely a consequence of the collective limitations of these treaties.[4]

The first half of the twentieth century marked the heyday of species-protection treaties, at least as measured by the sheer number of such treaties that were negotiated, signed, and ratified. But these treaties are only explicable within the context of the scientific-technological revolution (or "second industrial revolution") that began in Europe and North America during the latter half of the nineteenth century. The ever-increasing demand for coal, petroleum, lead, copper, tin, ivory, rubber, coffee, bananas, tropical oils, and hundreds of other natural resources spawned a great amount of competition among the European powers, pushing them in the direction of global imperialism. So did the construction of railroads and canals, the development of steel-hulled ships, and the invention of nitroglycerin. The result was a "scramble for Africa" that brought the sub-Saharan regions almost completely under European domination, a land rush that turned much of the western United States and Canada into cities and irrigated farmland, and a mad dash to Antarctica by the largest whaling companies. New killing techniques played a major role as well. One thinks here especially of the breech-loading and magazine rifles that Europeans took with them to Africa, the double-barreled shotguns that U.S. citizens pointed toward the skies of North America, and the grenade-tipped harpoon gun that Norwegian whalers used with such devastating effect on the high seas. Collectively, these forces initiated what can aptly be described as a war of extermination against the world's wildlife.

No doubt, subsistence and recreational hunters had a modest impact on the world's wildlife stocks in the early twentieth century, but it was market hunters who caused the bulk of the devastation. Market hunting was (and, on the high seas, remains to this day) essentially an extractive industry. Market hunters depleted species the way miners depleted ore seams, moving to new sites after exhausting the old ones, thinking only of today's profit and not tomorrow's patrimony. Behind the killing frenzy in Africa was the enormously lucrative trade in ivory, skins, and feathers, with ivory commerce alone accounting for most of the profit. Behind the avian slaughter in North America were the millineries and meatpacking industries, which turned millions of birds each year into hats and meat. Behind the boom in whale hunting was the demand for edible fats, with millions of pounds of blubber ending up as lard and margarine in the kitchens of Europe. The wastage was phenomenal. Elephant hunters took only the tusks, leaving the carcasses to the buzzards. Bird hunters would sometimes wipe out entire rookeries and flocks in a single day, with little thought to future migrations. And whalers could lose as much as one-fourth of their catch to the treacherous waters of the Antarctic. What made Leopold's *Game Management* so important was that it called for a more sensible model of wildlife conservation, one that replaced the mining mentality of the market hunter with the more sustainable model of farming. The goal of farmers is not depletion but maximum yield.

Today, it is relatively easy to distinguish a market hunter from a sport hunter or a subsistence hunter, but a hundred years ago, the lines were still a bit blurry. Sport attracted thousands of Europeans and Americans to Africa on safari (Arabic and Swahili for "journey" or "caravan") each year, but few of these hunters showed any qualms about recouping part or all of their travel costs by selling tusks and other animal products on the open market. Similarly, a western settler in the United States or Canada might kill one bird species for subsistence, another for sport, and a third for the market—sometimes all on the same day. Whaling was more recognizably divided into market and subsistence hunting (sport hunting was all but nonexistent), but even here, there were crossovers: subsistence hunters often sold or traded what they did not consume, either to nearby communities or to faraway markets. By helping to establish different regulations for market, sport, and subsistence hunting, the treaties discussed in this book played a modest role in the creation of more distinct lines. And by promoting recreational over market and subsistence endeavors, they also helped create a hunting hierarchy.

Three words in the title of this book require clarification: *game, conservation,* and *migratory.* The term *game* derives etymologically from *gaman,* Old High German for "amusement," a connotation that it still carries today. Any activity engaged in for pleasure or diversion—from professional soccer to church bingo—can be considered a game so long as the players adhere to an agreed-upon set of rules. Games are almost always associated with a certain amount of levity and frivolity, even if the players take their pursuit with utter seriousness. Only later did the word *game* become associated with hunting. No doubt this newer usage evolved from the amusement that European aristocrats derived from sport hunting, but *game* nowadays refers to any animal that humans hunt on a regular basis, whether for pleasure or for subsistence.

One of the chief purposes of any hunting law or treaty is to spell out the "rules of the game" in order to ensure that the prey remain plentiful for future generations of chasers. Fowl-hunting regulations in North America offered a success story in that regard. In other cases, however, laws and treaties turned out to be "games" in a different sense: they were diplomatic "diversions" that provided a legal framework behind which the carnage continued. Elephant and rhino hunting in Africa is one prominent example, and pelagic whaling in the Antarctic another, for no international agreement has ever successfully curtailed the hunt for ivory tusks, rhino horns, blubber, and whale meat. The game metaphor can certainly be carried too far, but surely it is worth noting that up to the mid-twentieth century, a large portion of ivory tusks were turned into billiard balls (and what is more frivolous than a game of pool?); that one of the principal motives for killing birds all over the world was to obtain their plumage (and what is more frivolous than fashion?); and that whalers often jokingly referred to their annual sojourn to the Antarctic krill grounds as the "whaling Olympics" (and what is more frivolous than to declare the enterprise that massacred the most whales in the shortest period of time as the winner?).[5]

Game is a highly problematic term, for there has never been universal agreement as to which animals should be targets and which should not. Most hunters and conservationists would probably agree that antelopes and ducks are game, but what about robins? New England bird-watchers always tended to see them as beautiful songbirds, but in the impoverished southern United States, they were once considered the main ingredient in "robin soup." Captains of industry could readily defend the use of elephant tusks and whale baleen in a wide variety of commodities before the invention of plastics, but what about afterward? The demand for ivory and baleen continued long after there were readily available substitutes. Similarly, what genuine economic justification was there for the enormously expensive annual expeditions to the Antarctic, when many common plants—including palm, coconut, and flax (linseed)—produced edible oils that were all but identical to blubber oil? Did whales have to face near extermination so that there would be one more oil source on the world market? The answers to these questions are more political than ecological: the powerful decided what was "fair game"; the powerless did not.

Game also has a problematic antonym—*vermin*—defined as any animal that deserves eradication because it competes in some way with the spread of human settlements or agricultural growth. This term has some elasticity (elephants are game when hunted for their tusks but vermin when they trample crops), but there was always a good deal of consensus about which animals were meant: predators, such as crocodiles, lions, bears, wolves, and coyotes, that competed with humans for the same game animals or fed on domestic herds. If *game* and *vermin* have come to have an old-fashioned ring today, this is largely because the term *wildlife* (originally *wild life*) gradually supplanted them. But in the early decades of the twentieth century, *game* and *vermin* were more commonly invoked in everyday speech and diplomatic discourse than was *wildlife;* moreover, there was a much greater tendency to see animals as good or bad based on their behavior rather than in terms of their contribution to a stable ecosystem. (Who today divides feathered species into "game birds," "birds useful to agriculture," and "crop pests"?)

The term *game law* also has an antithesis—*lawlessness,* better known as *poaching* and (for whaling) *pirating.* Latin law, upon which both the

European and U.S. legal systems were based, viewed a wild animal as *res nullius,* an entity that belonged to no one until it was captured or killed. When European and U.S. governments later decided to regulate free-roaming animals, they were in effect asserting some level of proprietary rights over these animals while they were within their territory, much in the same way that property owners typically think of animals on their land as theirs. Enforcement, however, is no easy matter, so the effectiveness of game laws depends in large part on the willingness of most citizens or subjects to obey them, as well as on the willingness of authorities to implement them with force. When there is widespread resentment to a law—or an easy way to evade it—it will prove ineffective. The battle between enforcers ("wardens") and resisters ("poachers") is a longstanding one, and it is a battle laden with class and ethnic conflict. The introduction of game laws in early modern Europe pitted peasants against aristocrats, subsistence hunters against sport hunters, and local officials against government regulators. When game laws were introduced in North America, many settlers resented the fact that their everyday activities, such as hunting to put meat on one's table, had suddenly become criminal acts. When Europeans later foisted game laws on their African colonies, the locals responded much as European peasants and American settlers had before them: they ignored and evaded the laws as best they could. "Every African is a poacher," Kenya's chief game warden, William Hale, would summarily pronounce in 1953, without a hint of irony or self-reflection.[6]

Environment and *ecology* are the buzzwords of wildlife protection today, but a century ago, the terms *preservation* and *conservation* reigned supreme, especially in the Anglo-American world. *Preservation* is typically associated with any effort to protect a specific species or a specific landscape from economic development or exploitation. This is often seen as a "hands-off" attitude to wilderness, but in practice, it was more of a "light-touch" approach, since most preservationists fully expected people to visit the protected sites and to use them for hiking, recreation, leisure, and touring. *Conservation,* by contrast, implies a commitment to the use of natural resources—animals, trees, water, land, minerals, and so forth—in a sustainable (typically dubbed "wise") manner. *Preservation* was the term of choice in the nineteenth century, but *conservation* began to supplant it in the United States and elsewhere during the presidency of Theodore Roosevelt (1901–9).

Whether a change in terminology brought with it a genuine transformation in attitudes from "hands-off" to "wise-use" is not all that clear even in the United States, where the terms were in widespread use. It is even less clear at the international level. The 1900 London Convention, for instance, was cast in the language of species preservation, but its chief purpose was to create a sustainable basis for the trade in ivory tusks, animal skins, and bird feathers, an idea more associated with conservationism. The 1933 London Convention, by contrast, was couched mostly in conservationist terminology, but its most lasting impact was to promote the establishment of nature parks and game reserves, one of the chief goals of the preservationists. Similarly, the bird treaties of 1916 and 1936 were long on species preservation and short on habitat protection—despite the wise-use rhetoric that dominated the thinking of U.S. legislators at the time. The whaling agreements of 1931, 1937, and 1946 also belie the notion that conservationist rhetoric always translates into wise use: these treaties offered no protection to individual species until they had already become "commercially extinct" (that is, too rare to hunt profitably), while at the same time sanctioning the overexploitation of still-plentiful whale species under the guise of sustainability. In this book, the terms *preservationist* and *conservationist* will largely be used interchangeably, much as they were in international discourse during the first half of the twentieth century.

All animals move about in search of food and shelter, but not all are considered *migratory*. Biologists reserve that term to describe species that move with the seasons in search of suitable habitat and sustenance. Africa's wet and dry seasons largely dictate the movement of elephants, zebras, wildebeests, and other game mammals, most famously on the savannas of East Africa (modern-day Kenya, Uganda, and Tanzania). In North America, heat and cold dictate the movement of swans, ducks, geese, and hundreds of other avian species. They "summer" in the northern latitudes of the United States and Canada and then "winter" (a well-entrenched misnomer, since they actually move southward to continue their summer) in the more equatorial latitudes of the southern United States, Mexico, Central America, and South America. This summer-winter pattern is also present on the high seas: whales typically feed in the Arctic and Antarctic regions when it is warmest there (July in the Arctic, January in Antarctica) and then head toward equatorial waters to breed when the poles become too cold.

Diplomats use a more restrictive definition of *migration* than do biologists: only those species that regularly cross national borders in their seasonal movements come under their purview, whereas those that stay within the confines of a single state remain dependent on domestic (or colonial) legislation for protection. This restrictive distinction is often inconsequential. Zebras, for instance, inhabit Tanzania's Serengeti National Park during the wet season (November to May) and then move to Kenya's Masai Mara National Reserve during the dry season (June to October). Similarly, the Canada goose oscillates between its namesake nation and Mexico each year, and the American golden plover cycles between northern Alaska and the tip of South America. The gray whale, meanwhile, travels nearly ten thousand miles between its feeding grounds in the Bering and Chukchi seas and its breeding grounds in Baja California, the longest known migration of any mammal in the world. Animals that migrate thousands of miles typically cross many borders on their journey, making them the legitimate subject matter of treaty making. Occasionally, however, the term *migratory* has given rise to diplomatic ambiguities and legal challenges. In *United States v. Lumpkin* (1921), for example, a federal judge in Georgia had to decide whether mourning doves were migratory and therefore subject to the terms of the 1916 U.S.-Canadian bird treaty or whether they were nonmigratory because some flocks never flew far enough north to reach Canada during the summer months. (The judge took a pragmatic position, ruling that mourning doves were migratory because the treaty said so!)

The treaties analyzed in this book cover a wide variety of species over a diverse range of animal habitat—on land, on sea, and in the air. The African treaties were quintessential colonial accords: they were written by European administrators, not African leaders, and they reflected the priorities of colonial officials, not the indigenous populations. On the positive side, these treaties attempted to rein in the export trade in animal products and to establish protected sites for females and their young. On the negative side, they turned black Africans into poachers and permitted the removal of the Masai and other groups from territories designated as nature parks. The North American bird treaties were written mostly by U.S. conservationists and ornithologists, and they reflected the needs of U.S. hunters more than Canadian and Mexican ones. On the positive side, they helped bring an end to the commercial market in food and feathers, which was wreaking havoc on avian species worldwide at the beginning of the twentieth century. On the negative side, they did little to preserve habitat along North America's four great migration routes (known as the Atlantic, Mississippi, Central, and Pacific flyways), and as a result, many protected birds today have trouble finding adequate places to feed and rest on their journeys. The whaling treaties were conceived and written by the major whaling nations, and they were designed to protect the business of whaling. On the positive side, they offered protection to those species that had gone commercially extinct, thereby saving remnant populations of rights, bowheads, and grays from complete extermination. On the negative side, the International Whaling Commission (IWC)—the regulatory agency established in 1946 to protect the stocks for future generations—sat idly by as the major whaling companies brought many other whale species (blues, fins, and humpbacks, among others) to the brink of commercial extinction. A treaty that protects a species only after it has been decimated is hardly one that can be held up as a positive achievement.

Despite differences in scope and substance, the African, North American, and Antarctic treaties all had much in common. Each began as a national endeavor and then evolved into an international agreement. Germany (and later Great Britain) was the driving force behind the African treaties, the United States behind the bird treaties, and Norway behind the whaling treaties. Each treaty also originated as a piece of domestic (or colonial) legislation and then grew into something transnational. The German East African Game Ordinance of 1896—designed to regulate the trade in ivory, skins, and feathers in modern-day Tanzania—jump-started the establishment of Africa's nature parks and game reserves. U.S. legislation, most notably the Lacey Act of 1900 and the Weeks-McLean Law of 1913, provided the backdrop for the North American bird treaties. And the Norwegian Whaling Acts of 1929 and 1935 provided almost all of the verbiage for the 1931 and 1937 whaling agreements.

Each treaty also began as a purely utilitarian effort to maximize game stocks for the benefit of future generations of hunters and then evolved (for better or for worse) into a more all-encompassing conservationist treaty. Few diplomats would have predicted, in 1900, that nature parks and game reserves would one day become the backbone of African conservationism, let alone that these protected areas would attract millions of camera-toting tourists each year in search of Eden. Even fewer thought, in 1916, that the U.S.-Canadian bird treaty would set the tone for avian protection in North America for the rest of the century and even act as a spur for similar bilateral treaties between the United States and Japan (1972) and the United States and the Soviet Union (1976). And no one foresaw, in 1946, that the IWC—the lapdog of the major whaling nations—would one day be transformed into an antiwhaling institution, though in fact that is what happened in 1982.

As the earliest industrial nation and the largest colonial power, Great Britain played a central role in formulating most animal-protection treaties in the first half of the twentieth century. The British Colonial Office hosted both the 1900 and 1933 African conventions, and British conservationists took the lead in establishing many of Africa's most famous nature parks and reserves. The British Foreign Office represented the Dominion of Canada (which was still a semicolony) in the negotiations with the United States over the 1916 bird treaty. British diplomats also exerted an immense influence over the terms of the whaling treaties, in part because Britain was a major whaling power (second only to Norway) and in part because a British-based consortium, Unilever, enjoyed a virtual monopoly over the global whale-oil trade. The U.S. presence deserves to be highlighted as well. President Theodore Roosevelt was actively engaged in the movement for African conservation, even if his safari excesses were a matter of international consternation. The United States played the lead role in the 1916 and 1936 bird treaties, and it hosted the 1946 whaling conference.

The strong presence of British and U.S. diplomats in the treaty-making process meant that Anglo-American notions of animal preservation and conservation tended to emerge victorious. It also meant that Anglo-American nongovernmental organizations were able to influence the terms of the treaty to a large degree. In Africa, the principal organization was the Society for the Preservation of the Wild Fauna of the Empire (also known as the Fauna Society), created in 1903 to lobby for the creation of larger game reserves and stricter game laws. It was nominally independent, but virtually all of its founding members were prominent statesmen and colonial administrators (not to mention big-game hunters) who maintained close ties to the British Foreign and Colonial offices. In North America, there were a variety of organizations—including the American Ornithologists' Union, the National Association of Audubon Societies, and the American Game Protective and Propagation Association—pushing for greater international cooperation to protect birds. These organizations too were dominated by avid hunters and ex-hunters who were now trying to save the animals they had shot with such gusto a few years earlier. The first successful international organization—the International Union for the Conservation of Nature and Natural Resources (IUCN)—also had a strong British and U.S. presence; it was active in whaling issues in the 1950s and 1960s, though it became active too late to help formulate the treaties themselves.

The lobbying effort of these "penitent butchers" helps explain why so many countries around the globe were willing to sign and implement these treaties. But the prominence of hunters in the negotiating process also helps explain some of the treaties' inherent weaknesses. They were largely designed to establish uniform game regulations across national borders so as to provide a level playing field for hunters and to reduce the illegal transport of products across borders (via fencing and smuggling). They tended to focus almost exclusively on game animals to the neglect of other species. And all too often they paid inadequate attention to habitat protection. Africa's earliest nature parks and game reserves were placed in areas that were considered to be economically useless, with little or no thought given to the migration routes; as a result, they often provided only part-time protection to game animals. Similarly, North American game hunters and legislators often showed more interest in setting aside land as public and private shooting grounds than as bird preserves, an obviously self-defeating policy in the long haul. The whaling nations, meanwhile, were quite open about the fact that they were willing to accept almost any restriction on hunting except the two that made the most sense from the perspective of conservation: a species-by-species annual quota based on stock size and reproduction rates, and large sanctuaries in key feeding and breeding grounds.

The economic liberalism and political decentralism of the Anglo-American tradition further limited the efficacy of the treaties. As an example, the British government focused almost entirely on the export market, and it put its funds into game wardens and customs officials in Africa. It largely ignored the import trade, even though the London commodities market (where a large portion of the world's tusks, skins, and feathers were auctioned) was headquartered in the same city as the Foreign Office and Parliament. Similarly, states' rights advocates in the United States did everything they could to thwart the bird treaties on the grounds that they would augment the power of the federal government at the expense of the states; they had a misplaced faith in the willingness of the individual states to create a sufficient number of reserves on their own. Though states' righters were ultimately defeated, they did manage to thwart the establishment of the National Wildlife Refuge System for many decades. And by that time, farmers had drained many of North America's premier wetlands, the very sites that migratory birds depended upon for their sustenance. The U.S. government repeated this mistake at the international level: it balked at the prospect of putting real regulatory teeth into the 1946 whaling treaty, relying instead on the so-called free market (which was actually a sheltered market) and on the goodwill of the major whaling nations and whaling companies—a mistake that proved nearly fatal to animal conservation in the Antarctic.

Issues of sovereignty also played a key role in determining the relative efficacy of these treaties. The European colonial powers were willing and able to work out agreements with each other, but they had trouble convincing the Swahili Arabs who still controlled some of the trade routes that the 1900 and 1933 African treaties were worth paying attention to; further, they failed to bring Africa's two independent regions, Liberia and Abyssinia (until 1935), on board. Many corrupt traders and colonial officials (including game wardens) were therefore able to circumvent the treaty restrictions by utilizing Swahili Arab middlemen to smuggle ivory, skins, and feathers. Similarly, Canada, the United States, and Mexico were able to work out an amicable arrangement for protecting birds in North America, but they were never able to bring the Caribbean, Central American, or South American states into the fold. Bird species with migratory routes that included the Southern Hemisphere were therefore only protected during certain times of the year. The whaling treaties were even more problematic. Enforcement would have been much easier if one country had controlled all of Antarctica or if the waters around that continent were under the jurisdiction of the League of Nations or the United Nations. But Antarctica was a continent without a people or a government, and the oceans around it belonged to no one, so it was child's play for whaling enterprises (legitimate and pirate ones alike) to circumvent the restrictions.

Given the hurdles, perhaps the most remarkable aspect of these treaties is that they came into being at all and that they managed to place some restrictions on hunting, even if (as was the case with whales) they could not bring a complete halt to the slaughter. "To keep every cog and wheel," Aldo Leopold wrote in *Round River*, "is the first precaution of intelligent tinkering."[7] Judged by this standard, all of the treaties—even the whaling ones—can be judged modestly successful: no African land mammals, North American bird species, or Antarctic whales have gone extinct on their watch.

Chapter 1

Africa's Apartheid Parks

> The word "ivory" rang in the air, was whispered, was sighed.
> You would think they were praying to it.
>
> —Joseph Conrad, *Heart of Darkness*

HERMANN VON Wissmann was one of Germany's most renowned African explorers. A travel writer and big-game hunter, Wissmann was best known for having traversed the southern Congo basin on behalf of Leopold II, king of the Belgians, in the early 1880s. Chancellor Otto von Bismarck later asked him to govern German East Africa (which he did intermittently from 1888 to 1896), not least because he was skilled at suppressing colonial revolts. Outwardly, there was little about Wissmann's career that distinguished him from other African imperialists of his day—Richard Burton, John Speke, Cecil Rhodes, Frederick Courteney Selous, H. M. Stanley, and Carl Peters among them—except in one regard: he was the primary champion of an international conference that would result in the London Convention for the Preservation of Wild Animals, Birds, and Fish in Africa in 1900 (hereafter the 1900 London Convention).[1]

All of Africa's major colonial powers attended the conference and signed the 1900 London Convention: Great Britain, France, Portugal, Spain, Belgium, Germany, and Italy. The first four had colonized prior to the 1880s, mostly along the African coastline. Britain's historical strong-hold lay in the Cape Colony (the nucleus of today's Republic of South Africa), on the continent's southern tip. The Suez Canal, completed in 1869, also gave Britain a toehold in Egypt, though that region was still nominally under the Ottoman Empire's control. Algiers (Algeria) was France's most important colony, but some of the coastal towns and hinterlands of west-central Africa (the nucleus of Senegal, Gabon, and a few other regions) were also within its orbit. Portugal controlled Angola on the Atlantic side of the continent and Mozambique on the Indian Ocean side. The Spanish influence was largely limited to the Canary Islands and a couple of specks (Ceuta and Fernando Po) along the northern and western coastline.[2]

Belgium, Germany, and Italy were new colonial powers. The Belgian presence began in the mid-1880s when Leopold II took possession of the Congo Free State (later Zaire and now the Democratic Republic of the Congo), deep in the heart of Africa, and ruled it as his personal fiefdom. A brutal overlord even by European standards, Leopold II turned his Congo colony into a rubber-, copper-, and ivory-producing sweatshop until his death in 1909, when it was turned over to a much-embarrassed Belgian government to administer. Germany's presence in Africa also began in the mid-1880s, when Bismarck took possession of German East Africa (roughly Tanzania minus Zanzibar), German Southwest Africa (Namibia), Togo-land (Togo and part of Ghana), and the Cameroons (Cameroon and part of Nigeria). The Italians seized most of Somaliland (Somalia) in 1885.[3]

The colonization of Africa proceeded slowly until the 1880s, but once the parvenu powers arrived, European statesmen began meeting periodi-cally to settle their differences and forge common policies. At the Berlin Conference (1884–85), they established the rules of the game for future landgrabs, agreeing that colonizing governments had to take real posses-sion of the land they claimed with settlers and troops, not just take paper possession through fanciful maps and colorful flags. Following the prin-ciple of so-called free trade, they also banned import and transit duties in the colonized territories and set up the framework for future consultations among the colonial powers. Four years later, at the Brussels Conference (1889), they decided to stamp out the internal slave trade, which was still extant in Zanzibar and some other Arab-controlled regions of Africa. They also decided to restrict the types of firearms and ammunition that could be sold to black Africans between the twentieth parallel north and twentieth parallel south (roughly south of the Sahara and north of Boer territory) and to sanction the introduction of colonial gun licenses and big-game hunting restrictions. A few years later, the Congo Free State, France, and Portugal met separately to sign and ratify the Congo Basin Convention (1892), which created uniform export duties on elephant tusks in the regions under their control. The 1900 London conference was the last of these meetings to occur before the outbreak of World War I. It was designed to create uniform hunting ordinances throughout colonial Africa and to jump-start a network of nature parks and game reserves.

The Berlin Conference triggered a "scramble for Africa," a frenzied attempt by leaders of the various colonizing powers to lay claim to as much territory as possible before the other powers beat them to it. By 1900, most of sub-Saharan Africa (excepting Ethiopia and Liberia) was under European suzerainty. The Brussels Conference, meanwhile, made it easier for Europeans to suppress colonial revolts by depriving local Africans of access to modern European weaponry. It also had the inadvertent consequence of forcing African hunters to rely on "traditional" hunting techniques and equipment (spears, pits, traps, poisons, outdated muskets, and the like), while allowing Europeans to use "modern" ones (such as high-powered rifles, machine guns, modern ammunition, and scopes). The Congo Basin Convention, finally, created an economic bond among the Congo colonists, one based largely on the ivory trade. For decades thereafter, Belgium, France, and Portugal thwarted all efforts to curb the commerce in tusks.

On the positive side, these diplomatic agreements suggested that European colonists shared a common vision about economic development, natural-resource use, and conservation. On the negative side, the Europeans arrived as conquerors, not as equals, and they showed little understanding of, or sympathy with, African cultures and traditions. Contradictions abounded. The Europeans made it illegal for Africans to acquire modern weaponry—and then demonized them for using "primitive" hunting techniques. They usurped pastoral and agricultural space for their own cattle and fields—and then looked askance when Africans relied on wild animals ("bush meat") for their daily sustenance. They turned traditional hunting grounds into nature parks and game reserves—and then complained when Africans continued to hunt there. Policies that stigmatized the traditions of indigenous peoples under the banner of conservation and modernization were not ones that promised much compliance, at least not in the short run.

Today, the full spectrum of species that once roamed freely across Africa is preserved only in the continent's three-hundred-some national parks and natural reserves. Although these parks and reserves include only a small fraction of tropical Africa's landmass, they are nonetheless not

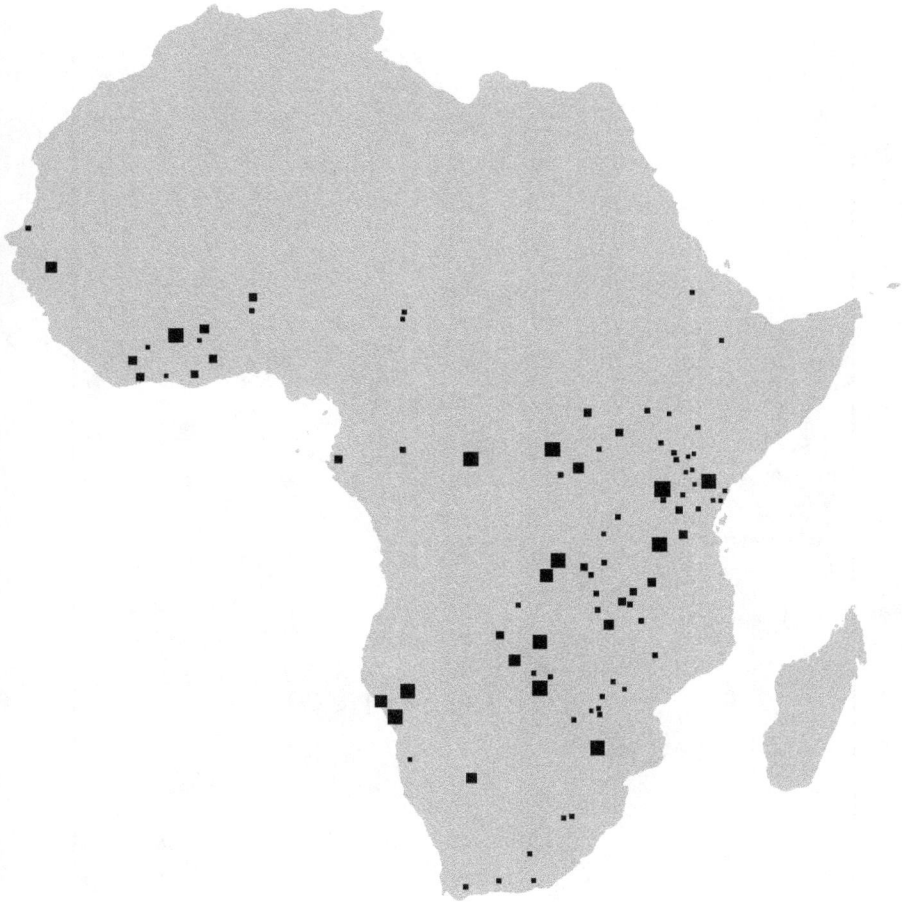

Figure 1.1. Map of Africa with major national parks and nature reserves. Adapted from Roland Oliver and Michael Crowder, eds., *Cambridge Encyclopedia of Africa* (Cambridge: Cambridge University Press, 1981), 313.

small in themselves, especially when compared to other protected regions in the world today.[4] The largest national park in the United States, Denali (in Alaska), is around 7,400 square miles in size. By contrast, Kruger National Park in South Africa is nearly 8,000 square miles; the Great Limpopo Transfrontier Park, which straddles South Africa, Mozambique, and Zimbabwe, is 13,500 square miles; and the Selous Game Reserve in Tanzania is over 19,000 square miles.

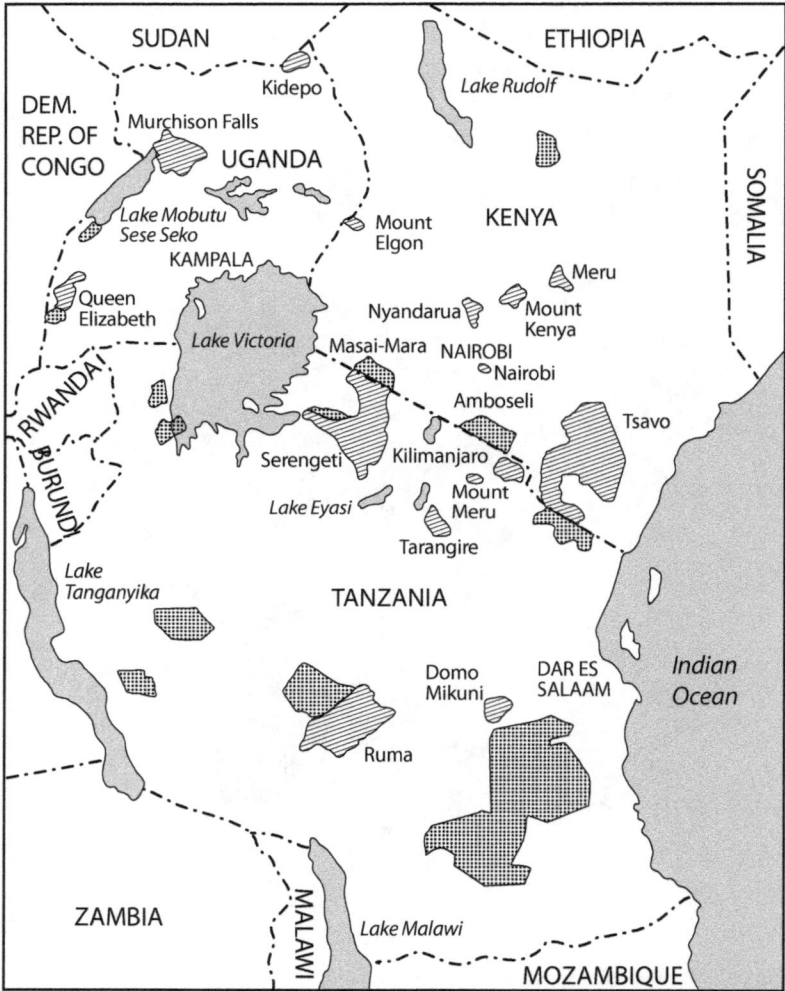

Kenya	Parks	9,200 km²
	Reserves	3,500 km²
Tanzania	Parks	12,000 km²
	Reserves	24,500 km²
Uganda	Parks	2,800 km²
	Reserves	200 km²

▨ National park
▦ Game reserve

Figure 1.2. National parks and nature reserves in East Africa (Kenya, Uganda, Tanzania). L. A. Lewis and L. Berry, *African Environments and Resources* (Boston: Unwin Hyman, 1988), 18. Reproduced by permission of Taylor and Francis.

These parks are a direct legacy of European colonial rule, and they cater mostly to European and American tourists, so it is not surprising that black Africans today often refer to them as "white man's parks." Equally apt, however, is the less frequently heard name "apartheid parks," for the parks were established along the same model used in the United States and elsewhere: indigenous populations were, for the most part, removed from the protected areas and new groups forbidden to migrate there, the only permanent inhabitants being animals and plants. The pecking order outside the parks was white settler, indigenous black African, and wild animal, with the indigenous populations being only slightly above the wildlife in the minds of many colonists. The pecking order inside the parks was tourist, animal, and indigenous black. Racism allowed the European colonists to view Africans as a part of the "natural" landscape and thus subject to the same brute-force relocations and control technologies they employed to subjugate the nonhuman world.

Just as Europeans carved up the African continent with little regard for its geographic, climatic, and faunal divisions (or its linguistic, ethnic, and traditional frontiers), they paid scant attention to the migratory patterns of African wildlife when they established these parks. European political and economic needs, not ecology, determined the border lines: an ideal nature park or game reserve, in the eyes of most colonial administrators, was one located on land that was deemed economically useless because it was disease infested, devoid of minerals and other resources, unsuitable for agriculture, or otherwise ill-adapted for white settlement. Few asked whether game animals were actually plentiful in these locations, whether there were sufficient food and water resources within the park boundaries, or even whether the areas were large enough to sustain the migratory patterns of the animals that were allegedly being protected. The end result was a hodgepodge of poorly placed, ill-designed "megazoos" that offered only part-time protection for migrating herds.

When Bernard Grzimek (director of the Frankfurt Zoo), for example, undertook the first comprehensive aerial survey of animal populations in Serengeti in the 1950s, he discovered that there was almost no congruity between the park's borders and animal-migration routes: at no time of the year were all of the Serengeti herds inside the park, but at certain times of the year, there were virtually none.[5] Similarly, the proximity of Nairobi Park to Kenya's capital city made it a favorite destination for tourists after it was created in the 1940s, but its small size (a mere forty square miles) made it wholly unsuitable for protecting Africa's mammals (though lions used the park as an entry point into the city suburbs, much to the consternation of the inhabitants!).[6] Other parks faced variations on these problems. For the Europeans to have created something more viable, they first would have had to remove their political and cultural blinders and create protected areas that were feasible from an ecological and cultural point of view. Yet at no time did the colonists seriously contemplate leaving vast tracts of African land to the local peoples and the indigenous animals or minimizing the impact of their own disruptive presence—even in the newly colonized regions of Africa where there were still few European settlers. A pecking order along the lines of indigenous African, indigenous wildlife, and white colonist did not conform to their racialized worldviews.

Ill-conceived or not, these national parks and natural reserves were created through a considerable amount of European diplomacy, and they remain to this day, as the *Cambridge Encyclopedia of Africa* (1981) has succinctly noted, "the backbone of nature conservation in Africa."[7] Wissmann was the first to champion an international hunting convention, but he was merely giving voice to what many other imperialists were themselves coming to realize: that Africa's animal herds migrated at will across the newly created frontiers of the British, French, Portuguese, German, Belgian, and Italian colonies; that Swahili Arab, Indian, and European traders bought and sold animal products throughout the continent, following the dictates of international commerce and not the requirements of sustainable game cropping; and that no individual colonial government could hope to regulate the trade in ivory, skins, and feathers by itself. Even Great Britain's leaders, who controlled the lion's share of African colonies and (by way of the London auction houses) much of the ivory trade, understood the limits of unilateralism. That is why they quickly seized the initiative from the Germans; hosted the two major conferences on wildlife conservation; and became the driving force behind countless African game ordinances, tsetse fly conferences, and wildlife-management projects in the first half of the twentieth century.

Africa's "Big Game"

European colonists had been fascinated by the broad spectrum of wild fauna that flourished in the forests and savannas of sub-Saharan Africa—elephants, rhinos, buffaloes, lions, leopards, giraffes, hippos, apes, baboons, and gazelles, to name but a few—long before Leopold II and others seized control of the continent's interior in the late nineteenth century. So large were their numbers and so great their variety that each new wave of immigrants tended to view the continent as a vast animal Eden, a realm shaped by nature rather than culture. But the supposedly pristine Africa that so many Europeans saw when they arrived on the continent was the Africa of myth: the continent was, in reality, a cultural landscape, a terrain shaped and reshaped over millennia by human agency. Indigenous Africans constantly transformed the ecosystems through their daily activities, most especially through cattle keeping, agriculture, and fire setting. Fire was used to destroy tsetse fly habitat, to thwart forest growth, to clear pastureland, and to promote the spread of the game-rich savannas. Many of the grass-filled plains that Europeans mistook for natural were in fact culturally produced landscapes, game-cropping regions created and maintained by an annual fire regime.[8]

Yet Africa's animal populations were so hearty and the continent's terrain was so varied and spectacular that successive generations of European intruders could easily convince themselves they had arrived in a pristine place. And the sense of cultural superiority they carried with them was so strong and their prejudices against the indigenous Africans were so deep that they readily overlooked the role of human agency in the regions they encountered. When former president Theodore Roosevelt visited Africa in 1909 on safari, he saw a landscape awash in nature but not teeming with people:

> In these greatest of the world's great hunting grounds there are mountain peaks whose snows are dazzling under the equatorial sun; swamps where the slime oozes and bubbles and festers in the steaming heat; lakes like seas; skies that burn above deserts where the iron desolation is shrouded from view by the wavering mockery of the mirage; vast grassy plains where palms and thorn-trees fringe the dwindling streams; mighty rivers rushing out of the heart of the continent through the sadness of endless marshes; forests of gorgeous beauty, where death broods in the dark and silent depths.[9]

Similarly, when Sir Julian Huxley went to Africa for the first time in 1929, he saw "a continent which had hardly changed in the last five hundred years."[10] And as late as the 1950s, Bernard Grzimek would claim: "Africa belongs to all who take comfort from the thought that there are still wild animals and virgin lands on earth."[11]

Colonists tended to classify Africa's mammals through the hierarchy of the hunt. At the top were the so-called trophy animals, most importantly the "big five": the lion, leopard, elephant, rhino, and (Cape) buffalo, all highly

YEAR	Belgian Congo	French Congo	Portuguese Congo
	WEIGHT Kilograms	WEIGHT Kilograms	WEIGHT Kilograms
1888	54,812	18,658	28,276
1889	113,532	3,601	9,284
1890	180,605	6,305	9,412
1891	141,775	16,661	7,469
1892	186,521	4,815	3,205
1893	185,993	3,142	1,287
1894	252,083	7,238	1,192
1895	292,232	24,381	101
1896	191,316	50,738	464
1897	245,824	53,908	477
1898	215,963	84,402	254
1899	291,731	78,800	103
1900	262,665	118,434	117
1901			
1902	249,307	133,491	160
1903			
1904	166,948	132,406	21
1905	211,338	152,986	40
1906	178,207	131,424	74
1907	203,583	143,355	369
1908	228,757	138,345	411
1909	243,823	135,237	405

Figure 1.3. Ivory exports from the Congo colonies between 1888 and 1909. Total annual exports grew nearly every year, from 101,746 kilograms in 1888 to 379,465 kilograms in 1909. George Frederick Kunz, *Ivory and the Elephant in Art, in Archaeology, and in Science* (Garden City, NY: Doubleday, 1916), 463.

prized because they were rare, elusive, or dangerous. Other trophy animals included the zebra, giraffe, and eland. Though not as difficult to hunt as the big five, they were still prized for their skins, antlers, or heads. Below them were the "pot" animals—most notably, the smaller antelopes—that seemed ready-made for shooting, if more for the meat than for the accolades. At the bottom were the "vermin"—a group that included the baboon, wild dog, and hyena, as well as the lion and leopard (two of the big five). Many colonists favored their wholesale extermination because they fed on domesticated livestock or competed with hunters for the same game, even if it meant that the big five would one day become the big three.[12]

Africa's mammals can be more scientifically divided into three broad groups: primates, carnivores, and ungulates (hoofed animals). Africa's primates (humans aside) include the ape, monkey, bush baby, and lemur, most of which prefer the continent's tropical rain forests and mountains to its open savannas. Only the gorilla and the chimpanzee (both apes) were prized targets in the early twentieth century, and they were also the only primates to receive some protective status in the African treaties of 1900 and 1933. The second group, carnivores, are (as their name suggests) meat-eating predators that live on other animals. Not surprisingly, the list of Africa's carnivores—mongoose, hyena, leopard, lion, cheetah, fox, jackal, wild dog, weasel, and otter—is all but identical with the category of so-called vermin mammals. Few of these animals received any sympathy from colonial settlers—or, for that matter, protection from turn-of-the-century conservationists.[13]

The third broad group—ungulates—can be subdivided into ruminants and nonruminants. Ruminants are even-toed ungulates that feed on plant tissues and fibers, and they include the antelope, buffalo, and giraffe. Antelope (hollow-horned members of the Bovidae family) are by far the most plentiful ruminants; they come in a wide variety of sizes and shapes, including the duiker, steenbok, gazelle, springbok, reedbuck, waterbuck, rhebok, roan, sable, oryx, hartebeest, topi, blesbok, bontebok, wildebeest (gnu), impala, bushback, kudu, bongo, and eland. Ruminants tend to be niche-specific: they have a highly specialized diet (such as the leaf of a specific tree species) that limits their breeding range. But their general preference for grasslands and their proclivity to run in groups and herds for safety make them one of the most common animals on the African savannas. The nonruminant ungulates are older (in evolutionary terms) than the ruminants and also better adapted to eating a broader variety of vegetation and tolerating a wider variety of habitats. They include some of the most coveted trophy animals—the hippo, rhino, zebra, and elephant (a near ungulate)—as well as some less desirable ones, such as the bushpig and warthog.[14]

No animal is more identified with the continent than the African elephant (*Loxodonta africana*), the world's largest land mammal and also the world's main source of ivory. An herbivore, it uses its tusks—elongated teeth that continue to grow throughout its lifetime at a rate of nearly one

	Antwerp	London
1886		340,000
1887		330,000
1888	6,400	373,000
1889	46,600	301,000
1890	77,500	357,000
1891	59,500	421,000
1892	118,000	396,900
1893	224,000	359,000
1894	186,000	376,000
1895	274,000	344,000
1896	265,700	284,000
1897	281,000	278,000
1898	205,300	300,000
1899	292,500	267,000
1900	336,000	320,000
1901	312,000	288,000
1902	322,000	269,000
1903	356,000	224,000
1904	329,000	212,000
1905	339,500	245,500
1906	303,800	208,500
1907	312,400	241,000
1908	227,700	214,000
1909	377,000	310,000
1910	336,500	257,500
1911	342,413	246,000
1912	385,330	245,000
1913	454,776	236,250

Figure 1.4. Weight in kilograms of ivory auctioned at Antwerp and London from 1886 to 1913. George Frederick Kunz, *Ivory and the Elephant in Art, in Archaeology, and in Science* (Garden City, NY: Doubleday, 1916), 467.

pound per year—to grub roots and to strip bark from trees. Elephants are famous for bulldozing their way through forest and brush (not to mention plantations and fields) as they migrate long distances in search of food and water. Their natural range includes all of sub-Saharan Africa, though hunting and habitat loss have severely restricted their movements today.

There are two subspecies of African elephant. The bush elephant is the larger of the two: it can attain a weight well over ten thousand pounds and reach a height of eleven feet. Its tusks, especially when small, are "soft," making them ideal for carving. Mature tusks become long and heavy (in the 1890s, an average tusk in East Africa weighed nearly sixty pounds).[15] The forest elephant has a round ear and is smaller than a bush elephant, averaging around seven thousand pounds and attaining a height of nine feet. Its tusks are also shorter and characteristically "harder" (more brittle). Its range is largely confined to the Congo basin and West Africa, though it can be found as far east as Uganda.[16]

Elephant herds are matriarchal. The oldest breeding cow is the leader, and the herd has around ten members, mostly calves, adolescents, and adult daughters. The twenty-two-month gestation period is unusually long for a mammal, as is the two-year suckling period. A cow can produce ten or more children across a lifetime that averages fifty to sixty years, so many females spend a considerable amount of their adulthood gestating or lactating. Females stay with the herd for life, but males strike off on their own upon reaching puberty at twelve to fifteen years, though they may join other herds for periods of time. When the matriarch is injured or killed, the rest of the herd is reluctant to abandon her, an instinct that often proved fatal to the entire herd in an era when get-rich-quick ivory predators roamed the continent.[17]

Before the widespread use of high-powered rifles and scopes, elephant hunting was a hazardous enterprise. Many Africans simply left these animals alone and sought their protein from more easily procured sources, but some groups specialized in killing elephants for subsistence and (once there was a thriving export market) for profit. The main obstacles to a successful hunt were the elephant's intimidating size, sharp tusks, good sense of smell, and thick hide, which, taken together, made it difficult to get close enough to land a deadly blow. The Waata (of Kenya) overcame these hurdles by using a powerful poison derived from the Acocanthera tree. Hunters placed the poison on the tips of their arrows, covered themselves with elephant dung to disguise their smell, and then crept into the herd before taking their shots. Once an arrow pierced the hide, the poison would flow into the intestinal cavity, inducing cardiac arrest almost instantly. The

YEAR	WEIGHT IN POUNDS	VALUE IN DOLLARS	AVERAGE VALUE PER POUND
1896	1,091,100	$2,226,804	$2.04
1897	1,028,800	2,048,863	1.99
1898	1,000,200	1,990,838	1.99
1899	993,900	1,959,499	1.96
1900	988,900	1,933,334	1.95
1901	882,500	1,628,160	1.84
1902	1,082,100	1,931,332	1.78
1903	924,100	1,648,438	1.78
1904	904,500	1,754,293	1.94
1905	1,055,000	2,058,469	1.95
1906	985,500	1,978,042	2.01
1907	1,078,700	2,718,693	2.52
1908	934,500	2,094,700	2.24
1909	1,155,500	2,590,215	2.24
1910	1,120,000	2,397,833	2.14
Totals	**15,225,300**	**$30,959,513**	**$2.03**

Figure 1.5. Total imports of animal ivory into Great Britain from 1896 to 1910. Belgium, France, and Egypt were Britain's major suppliers. George Frederick Kunz, *Ivory and the Elephant in Art, in Archaeology, and in Science* (Garden City, NY: Doubleday, 1916), 455.

Nyoro (of Uganda) used a different tactic: they attached a trip wire to a large and heavy iron spear, which they positioned along an elephant trail; once the elephant's foot struck the wire, the spear would plunge into its neck or spinal cord. Other groups used rope traps, designed to lasso a leg. Once the lasso brought the animal to a standstill, it could be speared or hacked to death. Still other groups employed a pitfall, a deep pit lined with upward-pointing spears. Setting a fire to induce a directed stampede was an indiscriminate but often effective method as well. Many Africans also acquired firearms and ammunition, typically in exchange for ivory; for the most part, however, these were old-fashioned muzzle-loaders, often used in conjunction with, rather than instead of, traditional methods.[18]

The Europeans did not create a new trade in African ivory in the second half of the nineteenth century: rather, they usurped the already existing trade that Swahili Arab and Indian merchants pursued with local African groups. The principal trading center was the island of Zanzibar (now part of Tanzania) in the Indian Ocean, which was then under the control of the sultans of the Omani dynasty. The island specialized in three interrelated "products": slaves, ivory, and cloves. The slave trade itself peaked in the 1860s and then gradually declined, but Swahili Arab and Indian merchants continued to rely on slave labor to transport tusks from Africa's interior to Zanzibar and also to work the clove plantations. Hamed bin Muhammad el Murjebi, better known by the nickname "Tippu Tip" (meaning "The Sound of Guns"), was the most famous of these Swahili Arab merchants, but there were hundreds of lesser-known figures with equally menacing nicknames ("The Locust," "The Oppressor," "The Finisher") who plied in "black and white ivory" alongside him. More often than not, the Europeans who opened up Africa's interior to exploitation were merely following the slave-and-ivory routes that Swahili Arab and Indian traders had long ago blazed (Richard Burton, John Speke, David Livingstone, H. M. Stanley, and E. Lovett Cameron all launched their expeditions from Zanzibar). At first, the Swahili Arab and Indian traders found plenty of ivory in the coastal regions, but as hunters decimated one herd after the next, the ivory-and-slave routes began to stretch deep into the interior. The first merchants reached Lake Tanganyika in the mid-1820s, Uganda a few years later, and the Lualaba River (Upper Congo basin) a few years after that, creating three intermediary ivory marts at Nyangwe, Ujiji, and Kazeh along the way.[19]

If Zanzibar was controlled by Arabs, the global ivory trade was handled mostly by Indian merchants. They shipped tusks from Zanzibar to Bombay (now Mumbai), at that time the world's ivory entrepôt, and from there to other parts of India or to China, Europe, and the United States. In India, the hollow middle part of the tusk (known as bamboo ivory) was highly valued, as it was ideal for the production of Hindu wedding bracelets (bangles). In Europe and the United States, the solid tip was most valuable, as it could be transformed into billiard balls. Ivory was also used around the world in the production of artistic figurines, as well as in the production of piano keys, knife handles, buttons, and other common items.

By the 1880s, European ivory traders—H. M. Stanley, Emin Pasha, and Alfred Swann among them—had begun to eclipse Tippu Tip and his fellow merchants. Tippu Tip lost his hunting grounds in the Congo to the Belgian king, in Tanganyika to Germany, and in Uganda to Great Britain, all during the "scramble." Even Zanzibar itself fell into British hands in 1890. To the victors went the spoils: Mombasa (in British-controlled Kenya) and Dar es Salaam (in German East Africa) replaced Zanzibar as ivory trading centers, just as London and Antwerp replaced Bombay as world auction sites.[20] From then on, Europeans arrived en masse to hunt in Africa, dreaming of ivory glory. They came in steel-hulled ships with plenty of cargo space for tusks; they built railroads to connect the ivory interior to the ivory ports; and they carried "elephant guns," rifles so powerful that one well-placed shot to the head or the heart sufficed to bring an animal down (though "bang bang" shooters far outnumbered crack shots).

Many observers assumed the end was near. "The question of the disappearance of the elephant here and throughout Africa is, as everyone knows, only a question of time," Henry Drummond, the author of *Tropical Africa*, lamented in 1889: "The African elephant has never been successfully tamed, and is therefore a failure as a source of energy. As a source of ivory, on the other hand, he has been but too great a success."[21] Carl Schilling, author of the popular *With Flashlight and Rifle*, made a similar declaration in 1905: "The day is not far distant when it will be asked, '*Quid novi ex Africa?*' [What's the news from Africa?] And the reply will be, 'The last African elephant has been killed.'"[22]

That their predictions did not come true—in regard to elephants or any other big game—was largely due to the willingness of European statesmen and conservationists to curb the slaughter before it was too late.

Wildlife Conservation before 1900

It is paradoxical that Africa's modern conservation movement began with the European scramble for Africa, for the white colonists were more destructive on a larger scale over a shorter period of time than any group that preceded them. The southern African experience offered a foretaste of things to come. The Dutch Boers (meaning "farmers") who founded Cape Town in 1652 named the regions they settled after the animals they found there: Elandsberg, Rhenoster, Oliphant's River, Quagga Fontein, Gemsbok, Leeuw Spruit, and the like.[23] By the time Britain took control of the Cape Colony in 1814 (and the Boers trekked to the Natal, Transvaal, and Orange Free State in the 1830s), the place-names, as the missionary-explorer David Livingstone later observed, were already beginning to become but cruel reminders of the eland, rhinos, elephants, quagga, gemsbok, and lions that had once abounded there. Most of these animals could still be found in southern Africa, albeit in greatly diminished numbers, but not all of them had survived the colonists' assault. The blaauwbok (*Hippotragus leucophaeus*), an antelope with a blue velvet coat, was last seen at the end of the eighteenth century; indigenous to southern Africa, it was a favorite target for hunters before it went extinct. The quagga (*Equus quagga quagga*), a subspecies of the plains zebra found only in southern Africa, died out in 1883; its demise came at the hands of farmers, who saw it as a competitor to their sheep and who turned its hide into grain sacks. Meanwhile, two once-common animals were well on their way to becoming regionally extinct: the Cape lion (*Felis leo melanochaitus*), which could no longer be found south of the Orange River by the end of the nineteenth century, and the Southern Burchell's zebra (*Equus burchelli*), which disappeared from southern Africa early in the twentieth century. Other species, notably the Cape mountain zebra (*Equus zebra*), the bontebok (*Damaliscus pygargus*), and the white-tailed gnu (*Connochaetes gnu*), were endangered. The South African wild ostrich (*Struthio camelus*)—hunted for its feathers and eggs—would have been endangered but for the fact that the local colonists domesticated it for commercial profit.

Like most farmers and pastoralists, the colonists had an ambivalent relationship to the local wildlife. On the one hand, they did everything they could to eradicate these creatures from areas under cultivation ("cleaning" the land, in Boer terminology). Wild animals were deemed nuisances or vermin: lions and leopards ate sheep, elephants and zebras marauded crops, and antelopes and gazelles competed for grazing space with cattle. On the other hand, many of these same farmers also depended on the consumption or sale of elephant ivory, rhino horn, hippo teeth, ostrich feathers, and antelope meat for their economic well-being, and so, the prospect of complete extinction was cause for alarm. To ensure a modicum of sustainability, the Cape Colony introduced game legislation (a closed season, protection for immature animals, antitrespassing measures) for the elephant, hippopotamus, and bontebok in 1822. In the Transvaal, the first game-protection measure came in 1846, and more legislation followed in 1858, 1891, and 1894.[24]

For the most part, these early game laws served the colonists poorly as conservation measures: they protected only those species that were important for subsistence hunting ("for the pot," in the language of the day) or had a well-recognized commercial value, while at the same time singling out any species that trampled crops or preyed on game for complete eradication. They did, however, accomplish one thing: they fostered a network of early game reserves (some of which later became national parks), which offered a measure of protection for a wide variety of species that were neither game nor vermin. The Natal established the Hluhluew, St. Lucia, and Umfolozi reserves in 1897. The Transvaal established the Pongola Game Reserve in 1894 and the Sabi Reserve in 1898. A Volksraad proclamation of 1895, which gave rise to the Sabi Reserve, told the tale:

> The undersigned, seeing that nearly all big game in this Republic have been exterminated, and that those animals still remaining are becoming less day by day, so that there is a danger of their becoming altogether extinct in the near future, request to be permitted . . . to discuss the desirability of authorizing the Government to proclaim as a Government Game Reserve, where killing of game shall altogether be prohibited, certain portions of the district of Lydenburg, being Government land, where most of the big game species are still to be found, to wit the territory situated between the Crocodile and the Sabi Rivers.[25]

Ironically, the crocodiles in the Crocodile River received no protection. They were "vermin" and therefore worthy of eradication.

South African colonists managed to wreak much of their eco-havoc with the use of inaccurate and slow-firing muzzle rifles. By the time a new influx of Europeans colonized Africa in the 1880s, breech-loading and magazine rifles were in common use. These were high-velocity weapons powerful enough, owing to their accuracy and rapid-fire capabilities, to obliterate entire herds in a short period of time. Moreover, though colonists still hunted primarily for meat and profit, the ritualized sporting enterprise— the safari—had come into vogue among Europe's privileged classes.[26] The most fashionable safari sites were British East Africa (Kenya) and German East Africa (Tanzania), though British Uganda and many other colonies saw an upswing in visitors as well. Safari enterprises, mostly headquartered in Nairobi and Dar es Salaam, proved enormously profitable and helped spur the colonial economy: a typical two-month safari in 1910 cost $1,700 (roughly equal to $30,000 today) and employed thirty or more black servants as headmen, gun bearers, cooks, butlers, horse boys, and porters.[27]

While rich Europeans went on safaris, the less well to do sojourned in Africa to prospect for gold and diamonds; to establish coffee, cotton, and banana plantations; or to pursue dozens of other commercial and industrial enterprises. The influx of whites in turn fueled a railroad-building craze reminiscent of that in the American and Canadian West a half century earlier. The Germans built the Central Railway from the port of Dar es Salaam to Kigoma (Lake Tanganyika), while the British built the Uganda Railway from the port of Mombasa to Kisumu (Lake Victoria). The railroads made it possible to establish new white settlement communities—in the Kenyan Highlands, for example—at the expense of game territory and black African hunting grounds. Railroads also made it possible for white hunters to enjoy an ersatz safari in the African interior. "Thanks to the Uganda Railroad, many government roads and bridges, and a network of well-defined native paths," Richard Tjader, a big-game hunter, wrote in 1910, "most parts of the country are now easily, comfortably, and safely reached, so that even ladies may greatly enjoy a short sojourn in the Protectorate."[28]

Europeans used terms such as the *white man's burden, mission civilisatrice* (civilizing mission), and *Kultur* (culture) to justify their presence, but local blacks (and medical historians) remember the years from 1890 to 1930 more as a time of epidemic disease rather than enlightenment and prosperity. Outbreaks of cholera and smallpox hit eastern Africa in the 1890s, decimating indigenous populations. Rinderpest, a cattle virus, as well as many lesser-known animal diseases, attacked domestic and wild herds at the same time. Many wild species were immune to rinderpest, but the buffalo, eland, kudu, and zebra were not, and they succumbed to the disease in large numbers. Cholera, smallpox, and rinderpest had immediate social consequences, for the pathogens sapped the military strength and economic viability of the Masai, Ngoni, and other African groups and starved the survivors into submission. "Through all this great plain we passed carcasses of buffalo," Frederick Lugard, one of Britain's most celebrated elephant hunters, wrote in his diaries as he traveled through Kavirondo, "and the vast herds of which I had heard, and which I hoped would feed my hungry men, were gone! The breath of the pestilence had destroyed them as utterly as the Westerners of Buffalo Bill and his crew and the corned-beef factories of Chicago have destroyed the bison of America."[29] It is hardly surprising, then, that the Masai used the word *Emutai* (derived from *a-mut,* meaning "to finish off") to describe their experience with this devastation.[30]

It is highly unlikely that any single East African species was in danger of extinction at the beginning of the twentieth century. The white population of East Africa was minuscule compared to that of South Africa. Kenya had just 600 white settlers in 1905, a number that would grow to just 9,651 settlers by 1921. As late as 1946, the white population was still under 25,000.[31] In 1911, Germany's four Africa colonies (East Africa, Southwest Africa, Cameroons, and Togoland) had a white population of 20,000 in a lebensraum of 1 million square miles.[32] Safaris, moreover, were an exclusive adventure of the rich, and even in the 1920s (the peak decade for their popularity), they accounted for only a small fraction of the total game killed. Though rinderpest decimated the buffalo and eland population in the 1890s, Uganda administrators felt compelled to remove buffalo from the endangered list a scant decade later, the herds having multiplied so fast that they had become "not only a constant nuisance, but also a serious danger to the people."[33] Nonetheless, Wissmann and many other colonial administrators understood that the cumulative impact of extravagant safaris, white settlements, and disease posed a long-term danger to the herds, and they moved the issue of African game conservation to the top of their colonial agendas.

The most important of the new colonial laws was the German East African Game Ordinance of 1896, which Wissmann wrote and promulgated. Though the measure was short-lived (Wissmann's successor as governor lifted it a couple of years later), it nevertheless served as a model for game laws and natural reserves elsewhere in Africa and also as a focal point for international negotiations. "I have too often seen how every European who possesses a gun on board a Congo steamer fires in the most reckless fashion, especially at hippopotami, without having any regard as to whether or not he can possess himself of the animal when killed," Wissmann wrote in 1897 to Baron Oswald Freiherr von Richthofen (head of the German Colonial Office from 1896 to 1898). "I have seen so much big game killed or mortally wounded in this wanton fashion, and, indeed, only by Europeans."[34] He convinced the Colonial Office that the time had come not just to restrict hunting but also to "turn some of the game-rich areas of German East Africa into a national park." Wissmann's model—despite a nominal nod to Yellowstone, the world's first national park—was more akin to a European hunting estate than a U.S. national park: he envisaged large game reserves that protected females and young during the critical breeding season, while still permitting the possibility of seasonal hunting for the privileged few. Wissmann himself picked the first two sites—one along the Rufiji River in the south and the other west of the famed Kilimanjaro—where hunting would be banned year-round, at least to most hunters, black and white alike. At the same time, he introduced a licensing system for the hunting of elephants and rhinos elsewhere in the colony, and he made it illegal to shoot elephants with tusks under three kilograms. The killing of females and young was prohibited year-round. He also outlawed certain hunting techniques, mostly indigenous ones (such as the use of fire to flush game), on the grounds that they were contrary to the "hunter's ethos." Many of these stipulations would find their way into the 1900 and 1933 treaties.[35]

Wissmann's role was important, but officials in other colonial states introduced measures on their own, both before and after him, and for much the same reason: they were concerned about the long-term viability of their game stocks, most especially the elephants. The Congo Free State promulgated an elephant-protection measure in 1889. German Southwest Africa gave protection to ostriches and other game in 1892. Hunting laws came to British East Africa (Kenya) in 1897, through the personal intervention of Lord Salisbury, Britain's prime minister and foreign secretary. "My attention," he told Commissioner Arthur Hardinge and Commissioner E. J. L. Berkeley of the East Africa and Uganda colonies, "has recently been called to the excessive destruction, by travellers and others in East Africa, of the larger wild animals generally known as 'big game.' There is reason to fear that unless some check is imposed upon the indiscriminate slaughter of these animals, they will, in the course of a few years, disappear from the British Protectorate."[36] In his response to Lord Salisbury, Hardinge divulged his sly strategy for game protection in British East Africa: "Keep as close as possible to the German Regulations, but make our own slightly more favourable to wealthy sportsmen who bring money into the territory."[37]

A colony-by-colony approach to game protection, however, proved difficult to implement in the absence of transborder cooperation. Thus, when German Southwest Africa banned the sale of female ostrich feathers in 1892, traders just smuggled their goods across the borders to Portuguese and British towns and ports, where they were able to exchange the feathers for ammunition and other goods. The Germans felt compelled to lift the ban a few years later and impose in its stead a nominal export duty of four German marks per kilogram on feathers.[38]

The need for uniform regulations was even more obvious in East Africa, where elephant, zebra, gnu, giraffe, eland, and other herds migrated between the frontiers of the German and British protectorates.[39] British game laws, for instance, mandated the confiscation of elephant tusks under five kilograms. Custom authorities, however, discovered the laws were impossible to enforce at ports such as Mombasa, Kanga, and Kismayu because traders could easily transport their contraband across the unguarded German and Italian frontiers without fear of confiscation.[40] When British authorities urged the Germans to adopt Britain's tusk-size regulations, the German authorities refused on purely practical grounds. "If the natives found that by crossing the Ruvuma they could find a ready market for their ivory," Paul Kayser (head of the German Colonial Office from 1890 to 1896) told Lord Salisbury, "all the ivory trade on the east coast would be diverted to the Portuguese possessions."[41] A few years later, when German East Africa finally did enact stringent tusk-size laws, Kayser's prediction came true: "Every prohibition to export," Governor Gustav Adolf von

Götzen lamented in 1903, "is an incentive to smuggling, and this will not cease so long as the Zanzibar market is not also closed to small ivory."[42]

If concern over the disappearance of East Africa's big game led the German and British colonial administrators in the direction of game laws, the logic of the export trade finally pushed them in the direction of international cooperation. When Wissmann floated the idea of an international agreement in 1897, Lord Salisbury not only concurred but also insisted that London host the conference (Wissmann wanted the conference to be held in Brussels). The British Foreign Office drafted a treaty and invited plenipotentiaries from Germany, Spain, France, Italy, Portugal, and Belgium. The conference commenced on April 24, 1900, and the delegates signed the London Convention on May 19.[43]

The 1900 London Conference: Environmental Laissez-Faire

The 1900 London Convention did not cover as much landmass as its grandiose title (Convention for the Preservation of Wild Animals, Birds, and Fish in Africa) might suggest. Article I delineated the twentieth parallel north (the same demarcation used at the 1889 Brussels Conference) as the northernmost limit of the treaty's jurisdiction. This demarcation line corresponded roughly to the faunal and political divisions that separated Saharan and sub-Saharan Africa, so it made sense from a conservationist perspective (though it did open up the potential for new smuggling routes along what was largely an unguarded border). More problematic was the conference organizers' decision to invite only the European colonists to meet in London, not the two remaining indigenous powers of Africa—Liberia, which had lost some territory during the "scramble" but had remained independent, and Abyssinia (Ethiopia), which had survived Italy's colonization bid in 1896. Both countries were located below the twentieth parallel north and were therefore within the purported jurisdiction of the treaty.

The southern demarcation zone, which followed the twentieth parallel south, made no sense whatsoever from the perspective of Africa's flora and fauna. This line was dictated solely by what the conference delegates delicately called the ongoing "difficulties" there, diplomatspeak for the Anglo-Boer War (1899–1902). It excluded a considerable amount of African territory, including all or parts of German Southwest Africa (Namibia), Portuguese East Africa (Mozambique), Bechuanaland (Botswana), Southern Rhodesia (Zimbabwe), and Madagascar, as well as the Cape Colony, Natal, Orange Free State, and Transvaal (Union of South Africa). After the war ended, the Germans, Portuguese, and British extended the zone southward to the Cape, but resistance to wildlife protection in the new Union of South Africa was evident for many decades thereafter.[44]

Article II and the schedules (which were appended to the treaty) spelled out the chief goals of the 1900 conference: to facilitate the creation of uniform game regulations by enumerating the animals to be protected, establishing closed seasons, and creating a licensing system. Section 7 of this article prohibited the "hunting of wild animals by any persons except holders of licenses issued by the Local Government." Since the vast majority of indigenous hunters could not afford these licenses (even if they were available for purchase deep in the African interior), this stipulation for all intents and purposes turned subsistence hunting into poaching. At the same time, Section 8 prohibited the "use of nets and pitfalls for taking animals," two trapping methods that Africans traditionally had used as means of subsistence and commercial hunting. (Subsequent game laws would ban still more traditional techniques, including foot snares, pits, traps, weighted harpoons, and poison-tipped arrows.) The putative reason for banning these techniques was that they were cruel to animals, but the actual reason was that the "passive" techniques of African hunters interfered with the "active" hunting methods of Europeans: horses fell into pits, and safari hunters got snared, trapped, and harpooned.[45] Meanwhile, the delegates reaffirmed the provisions of the Brussels Conference, which forbade the supply of modern arms and ammunition to African blacks. This too ensured that the onus of the treaty fell harder on the indigenous populations than on the Europeans. The convention deprived black Africans of their right to use traditional hunting methods without lifting the ban on the use of European weaponry.[46]

The schedules gave only a small number of species any real protection, nearly all of them central to sporting and commercial enterprises. Schedule I accorded full protection to eight animals "on account of their rarity and threatened extermination": the giraffe, gorilla, chimpanzee, mountain zebra, wild ass, white-tailed gnu (black wildebeest), eland, and Liberian (pygmy) hippo. Four birds were singled out for preservation "on account of their usefulness" to agriculture: the vulture, secretary bird, owl, and rhinoceros bird (oxpecker). Schedule II proscribed the killing of immature elephants, rhinos, hippos, zebras, buffaloes, ibexes, chevrotains, and various antelope and gazelle species. Schedule III prohibited "to a certain extent" the killing of females of these species "when accompanied by their young." Schedule IV set limits to the number of these animals (and a dozen or so others, including pigs, monkeys, cheetahs, and jackals) that could be hunted each year. Kill limits were also placed on several birds: the ostrich, marabou, egret, bustard, francolin, guinea-fowl, and "other 'Game' birds." In regard to fish, there was only one reference in the entire treaty: Article II, Section 9 prohibited "the use of dynamite or other explosives, and of poison, for the purpose of taking fish in rivers, streams, brooks, lakes, ponds, or lagoons."

Whereas Schedules I through IV *protected* specific species, Schedule V had the opposite goal of *eradicating* so-called vermin species: lions, leopards, hyenas, (wild) hunting dogs, otters, baboons and "other harmful Monkeys," large birds of prey (except the vulture, secretary bird, and owl), crocodiles, poisonous snakes, and pythons. The eggs of some animals—crocodiles, pythons, and poisonous snakes—were also singled out for destruction. The vermin clauses were designed to augment herbivore herds by controlling predators and to stop diseases (such as rinderpest) from jumping from wildlife to domestic herds. Wissmann, who attended the conference as an expert for the German government, advocated a policy of complete extermination, but Edwin Ray Lankester, director of the British Natural History Museum, called attention to the problematic nature of such a decision: "Certain species of animals, even if they are dangerous, should not be entirely exterminated because they are useful from other perspectives, such as preventing the excessive multiplication of other species." On Lankester's advice, the conference decided to permit a policy of animal control if it was "desirable for important administrative reasons," but it chose the phrase "reduce the numbers within sufficient limits" instead of "exterminate."[47] Still, it was odd that a preservationist document listed as vermin almost as many animals as it accorded the status of "full protection."

The debates over Schedules I through V proceeded smoothly, until discussion turned to the most lucrative area of African commerce—the trade in feathers, skins, and tusks. The British and German governments, in their original Draft of Suggested Bases for Deliberations of an International Conference for the Protection of Wild Animals, Birds, and Fishes in Africa (hereafter the British-German Draft) included a sweeping prohibition on "wholesale trade in the hides, horns, and tusks of wild animals and skins and plumage of birds."[48] The French and Portuguese, however, made it clear that they would not even attend the conference unless this clause was removed, so the British Foreign Office replaced it with a much milder restriction when it submitted its second draft, the Avant-Projet d'Acte Général (hereafter the Avant-Projet), at the conference itself. The reformulated text simply called for the "establishment of higher and more uniform tariffs for the exporting of the hides, skins, and tusks of wild animals, and the carcasses and feathers of birds."[49]

During the conference debates, the French and Portuguese delegates assaulted this reformulation as well, especially the phrase "higher and more uniform tariffs." The British delegates, for example, argued that the marabou stork should receive complete protection on the grounds that it was disappearing from the skies of West Africa ("It is a regular business in the Cassamance," one report read, "to give a native a gun with dust-shot cartridges and send him into the interior to shoot small birds for the milliners in Paris").[50] But the French representative, Louis Gustave Binger, compelled the removal of marabou from the fully protected list on the grounds that these birds were still plentiful in Senegal and "their feathers are an object of commerce."[51] The irony of the French position was not lost on the other plenipotentiaries: a conference whose avowed purpose was to stop "the destruction of animals for the purpose of pecuniary gain,"[52] it was noted, was scratching marabou off the full-protected list on the grounds that "their feathers were an object of commerce."

The French also fought hard to protect their ostrich-feather enterprises. When Wissmann suggested placing a duty on feather exports, Binger countered, "Such a scheme would have the effect of putting the French Colonies at a grave disadvantage vis-à-vis the Cape Colony, which could avoid establishing an export tax and where energetic measures have been taken for the preservation of ostriches. We are entirely committed to protecting the species but cannot accept an export tax on feathers."[53] A compromise was then agreed upon. Ostriches were added to Schedule IV (giving them partial protection) and a clause was added to Article II guaranteeing "the protection of the eggs of ostriches" from wanton predation, but no export duty was imposed on feathers. This compromise too was fraught with irony: the French refused to regulate the trade in *wild* ostrich feathers because it would benefit southern Africans, who had established a sustainable feather-farming industry based on *domesticated* ostriches.

The Portuguese and French delegates then joined forces against the proposal to impose an export tax on all animal hides and tusks. "The establishment of 'higher and more uniform tariffs' on hides exported from a part of Africa," claimed the Portuguese plenipotentiary Jayme Batalha-Reis, "could result in damage to existing European industries, for example Portuguese firms that rely on hides of African origin. The zone demarcated by Article I is only a part of the natural region where African animals should be protected. We might therefore be pressed to invoke Article II, section 9, in such a way that the trade be restricted in areas situated outside of this designated zone, where export tariffs are lower or absent." Binger agreed: "If we handicap commerce in African hides by the establishment of higher and more uniform tariffs, then hides of American provenance would have a competitive advantage. It would suffice, and also be more useful, if we just enumerate certain hides whose export would be forbidden."[54] In the end, the French and Portuguese prevailed. The final text imposed export duties only on certain specified products—"on the hides and skins of giraffes, antelopes, zebras, rhinoceroses, and hippopotami, on rhinoceros and antelope horns, and on hippopotamus tusks."

Conspicuously absent from this list was the elephant tusk! The Avant-Projet foresaw a high export tax on tusks that weighed less than five kilograms. The rate would increase ad valorem on tusks between five and fifteen kilograms. All tusks above fifteen kilograms would be subject to a uniform tax. But France, Portugal, and Belgium—the three parties to the 1892 Congo Basin Convention—closed ranks and refused to modify their existing arrangements. "We will not achieve important results from the point of view of animal protection by imposing tariffs, even high ones, on large elephant tusks, which will still have a considerable commercial value," Binger stated. "But since small tusks are less valuable, one could shut down legitimate commerce on them completely just by imposing heavy tariffs on them."[55] Bowing to the inevitable, Wissmann proposed a total ban on the export of tusks that weighed less than five kilograms but no export duty at all on heavier tusks. This compromise proved acceptable to a majority of the delegates. Article II, Section 11 of the convention thus simply proscribed "hunting or killing young elephants" and imposed "severe penalties against the hunters, and the confiscation in every case, by the Local Governments, of all elephant tusks weighing less than 5 kilogrammes." This was a small victory for the British, who placed special importance on the protection of immature elephants. But it was a major victory for the Belgians, French, and Portuguese, who were determined to defeat all efforts to impose export duties on mature ivory tusks.

Article III foresaw the establishment of game reserves within eighteen months of the treaty's ratification. The convention defined reserves as "sufficiently large tracts of land which have all the qualifications necessary as regards food, water, and, if possible, salt, for preserving birds or other wild animals, and for affording them the necessary quiet during the breeding time." Within the reserve, it was to be "unlawful to hunt, capture, or kill any bird or other wild animal," except vermin. The Avant-Projet text called for the establishment of animal sanctuaries in all colonies encompassed by the convention, but Binger reworded Article 2, Section 5 to read: "establishment, *as far as possible,* of reserves" within the convention zone. "The creation of Reserves does not appear to be viable in certain very populous colonies,

such as Gambia," he explained.[56] Batalha-Reis, the Portuguese plenipotentiary, expressed even greater hesitancy about nature reserves. He demanded
the assurance that local administrators would have the right to create reserves solely for the protection of specific species (such as the elephant or
zebra), the prerogative to alter the boundaries of the reserve as they saw
fit, and the authority to disband the reserves entirely if they so desired.[57]
All of these demands were duly written into the treaty text—a significant
watering down of the treaty, as it left the reserves hostage to the changing
whims of colonial administrators. Finally, the delegates inserted an elastic
clause into Article III to permit the suspension of the treaty's stipulations
as "necessitated by temporary difficulties in the administrative organization of certain territories."[58] This clause was designed to give administrators leeway during times of civil strife and disease epidemics, but of course,
it also provided them wiggle room for reneging on their promises.

The plenipotentiaries signed the 1900 London Convention at the conference's end, with the understanding that it would be valid for fifteen years,
after which it could be renewed, modified, or allowed to lapse. Problems
with ratification, however, arose even before the ink had dried. First and
foremost, the French representative announced at the concluding session
of the conference that France would not ratify the treaty unless Abyssinia
and Liberia also came on board. Neither of these countries, however, had
even been invited to the conference (and when later asked to sign, the
Liberian government blandly replied that its people "would most certainly
resent any attempt to prevent their shooting, or otherwise destroying, the
elephants which trample down their crops, or the leopards which carry off
their sheep and goats").[59] Leopold II's representative then announced that
Belgium would not ratify the convention unless France did so first. For its
part, Portugal held out until all of southern Africa was on board (which
would not occur until 1902).[60]

Ultimately, the refusal of the Congo powers to declare themselves whole-
heartedly in favor of the London Convention doomed its ratification. The
treaty bounced around for a dozen years or so, before being quietly shelved
when World War I broke out in 1914. Still, the conference partially achieved
two of its primary objectives. First, nearly all the colonial governments rewrote their game ordinances to conform to the principles laid down in
the convention, which included laws mandating closed seasons, minimum
tusk weights, licenses, and protection for immature animals and endangered species. Second, many governments at least made a nod in the direction of setting up protected areas, and several of them established extensive
networks of national parks and nature reserves.[61] To their credit, France,

Belgium, and Portugal were among those countries that rewrote their game laws and established protected areas in conformity with the convention. They did not, however, enforce the five-kilogram minimum on tusks or do much to regulate the trade in hides and feathers.

The British Experience, 1900 to 1933

The most immediate consequence of the 1900 London Convention was that it spurred the creation of the British-based Society for the Preservation of the Wild Fauna of the Empire. The Fauna Society, as its supporters affectionately called it, was founded in 1903 to lobby for the creation of larger game reserves and stricter game laws. For the next thirty years, it would be the single most important force for nature protection, not just in British colonies but in all of sub-Saharan Africa and much of Asia as well. It was nominally independent, but virtually all of its founding members were prominent statesmen and colonial administrators who maintained close ties to the British Foreign and Colonial offices. Because most of its members also happened to be big-game hunters or former hunters, detractors quickly dubbed them the penitent butchers.[62] It is an apt term, but it could be applied equally to nearly everyone involved with the 1900 London Convention or, for that matter, to nearly everyone involved in animal conservation in that era around the globe.

Efforts to establish similar lobbying groups on the continent met with some success, though none became anywhere near as powerful as the Fauna Society. The Wildschutzkommission der deutschen Kolonialgesellschaft (Game Protection Commission of the German Colonial Society), established in 1911, went defunct after World War I when Germany lost its African colonies to Britain and France. In Belgium, the Institut des Parcs Nationaux du Congo Belge (Institute of National Parks in the Belgian Congo) and the Fondation pour Favoriser l'Étude Scientifique des Parcs Nationaux du Congo Belge (Foundation to Promote Scientific Study of the National Parks of the Belgian Congo) were the driving forces in African affairs, and France's most important wildlife institution was the Société d'Acclimatation et de Protection de la Nature (Society for the Acclimation and Protection of Nature). Other European nationals founded similar organizations in their own countries, even if they possessed no African colonies.

Paul Sarasin (founder of the Swiss League for the Protection of Nature) and P. G. van Tienhoven (founder of the Netherlands Committee for International Nature Protection) attempted time and again to bring these national organizations under one roof, but they had only limited success.

The first International Congress for the Protection of Nature was held in Paris in 1909, and seventeen European nations signed the Act of Foundation of the Consultative Commission for the International Protection of Nature in Bern in 1913, but World War I broke out before the Consultative Commission ever had a chance to meet. After the war, the congress met twice in Paris (in 1923 and 1931), out of which the International Office for the Protection of Nature emerged in 1934. However, it did not survive the impact of World War II. It was not until 1948 that a viable international organization was established—the International Union for the Protection of Nature, in Morges, Switzerland (renamed the International Union for the Conservation of Nature and Natural Resources in 1956 and renamed anew the World Conservation Union in 1991). As a result, during the period between the two African conventions (1900 to 1933), the Fauna Society reigned supreme, as did British notions of nature protection.

The various national organizations spent much of their time lobbying for national parks and game reserves, and by 1933, they had achieved many successes. German East Africa (renamed Tanganyika Territory after the League of Nations placed it under British mandate) created eleven reserves, including the famous Serengeti, Kilimanjaro, and Selous reserves. The Union of South Africa established a total of eighteen protected areas, including the Etosha Game Reserve of former German Southwest Africa (which South Africa absorbed after World War I). The Anglo-Egyptian Sudan created six reserves, the Gold Coast one, Nigeria five, Nyasaland four, Northern Rhodesia five, Southern Rhodesia five, and Uganda six. British East Africa (renamed the Kenya Protectorate after World War I) possessed just two protected areas—the Northern and Southern reserves—but they were large, totaling around forty-eight thousand square miles together, about equal in size to the whole of England. The French established seventeen reserves in French West Africa, eleven in Algeria, ten in Madagascar, seven in French Equatorial Africa, and four in the Cameroons. Portugal created ten reserves in Angola and four in Mozambique. The Italian colonies had a total of eight. The Belgian Congo created thirteen protected areas, including the Parc National Albert (now Virunga National Park) in 1925, the first in Africa to be called a national park rather than a game reserve (though colonial park was more apt). South Africa followed suit, turning the Sabi Game Reserve in the Transvaal into the Kruger National Park in 1926 and adding three more parks in 1931.[63]

The British government took the lead in creating game departments to police the reserves and enforce the game laws, though personnel shortages and minuscule budgets hobbled all efforts at effective administration.

Revenues from game licenses and export duties contributed nearly 10 percent of the Kenya Protectorate's yearly budget (£68,069 in 1899–1900 and £121,692 in 1904–5), but only a small fraction of that money ever found its way to the Game Department, so the chief wardens were able to hire only a handful of assistants.[64] Under those circumstances, as one of Kenya's first game wardens, R. B. Woosnam, wryly observed, the new game laws had little impact on the behavior of hunters, "except that it gave birth to the ivory-smuggling trade."[65]

Enforcement improved over time, but even as late as 1939, the Game Department employed only five European officers and seventy African game scouts to patrol the entire perimeter of the Kenya Protectorate, which had a landmass of two hundred and twenty-five thousand square miles. The wardens, moreover, placed almost all of their personnel into the Southern Reserve, leaving the Northern Reserve to the Somali poachers who fenced tusks at nearby Italian-controlled ports. The port of Kismayu was a particular problem for the fencing of elephant ivory and rhino horn: "There is no possibility of suppressing the killing of the animal concerned so long as a free market exists over our borders," the Game Department concluded.[66] Assistant Game Warden K. F. T. Caldwell was even more blunt about the problems British officials were having with Italy: "Any specially protected animal can be immediately disposed of across the neighbouring border. Once such trophies have crossed the frontier they can be openly sold; *in fact to state that they were obtained from a neighbouring territory is accepted as a defence to any charge of their being illegally acquired.*"[67] In 1932, the British and Italian governments finally agreed to a joint effort to halt the poaching, but neither country sent enough game wardens and scouts to enforce the laws; when British troops seized Somaliland from Italy during World War II, they uncovered a still-booming business in illicit ivory and animal skins.[68]

British colonial administrators had to contend with discontent among indigenous Africans as well. "In Unyoro and Toro particularly, and in a less degree in other parts of the Protectorate, the Game Regulations have not been observed by the natives," the Uganda commissioner admitted to the Foreign Office in 1903:

> The business of procuring ivory is too lucrative not to tempt the Chiefs, and it is encouraged by the petty traders. . . . Before we took over the country, the necessities of the Chiefs and people, both in revenue and food, lay in the killing of elephants. This we have prohibited, giving them but little or nothing in re-

turn, and still expect them to be honest. All the Chiefs of Sazas get from the Government is 10 per cent on the cash which they collect for the hut tax, which in many cases does not amount to 100 rupees a year—too small a sum on which to keep up their position—whilst as regards the people at large, we have given them absolutely no return whatever.[69]

Six years later, the next Uganda governor wrote to the secretary of state for the colonies, this time to voice concerns about the newly created Toro and Bunyoro reserves. "The chiefs of Toro complained bitterly to me of the ravages of elephants, and begged for some relief," he wrote:

They asserted that the plantations were so frequently destroyed that the people are being forced to abandon the country. The elephants have become so fearless that they do not even hesitate to destroy habitations that stand in their way. They even attack travellers on the roads, and I was assured that, during the past year, no less than 16 persons have been killed by these animals in Toro alone. Under the game laws a peasant whose garden is being ravaged by elephants is not allowed to attempt to shoot them. He can only send to his chief, who is empowered to act in such cases, and is advised, in the meantime, to try to frighten the animals off by shouting and beating drums. The chief may take two or three days to reach the spot, and by the time he arrives on the scene the elephants are probably 30 or 40 miles off, and quite out of reach. The subject is one that bristles with difficulties, and while it would not be right to allow natives to kill elephants, save under exceptional circumstances, the fact remains that the animals are being protected to such a degree that they are devastating a populous and promising country.[70]

These complaints and dozens like them over the next decade prodded Uganda's colonial authorities to establish the Elephant Control Department (later renamed the Game Department) in 1923 and to embark upon extensive culling campaigns inside the reserves.[71] These efforts created the odd situation of having a game department that spent more of its time and money killing elephants than protecting them.

Yet another matter undermined all efforts to enforce colonial game laws and promote the establishment of protected areas: the tsetse fly problem.

At issue was whether the new game reserves were harboring tsetse flies and therefore promoting the spread of the disease trypanosomiasis. Trypanosomiasis is endemic in Africa and is better known as sleeping sickness when it infects humans and as *nagana* (a Zulu word) when it infects cattle. There are four organisms involved in the disease cycle: trypanosomes, which are flagellated protozoan blood parasites; wild animals, especially antelopes, buffaloes, warthogs, and bushpigs, which carry pathogenic trypanosomes in their bloodstreams but are immune to their effects; the tsetse fly (*Glossina*), a blood-sucking insect that feeds on large vertebrate animals; and a host, either a human being or a domestic animal. What made the tsetse fly central to this cycle was that it was the organism that transmitted trypanosomes from wild animals to people and domestic livestock through its bite. Different types of tsetse flies transmit different types of trypanosomes to different hosts, but only *Trypanosoma gambiense* and *Trypanosoma rhodesiense* are typically fatal to humans.[72]

Scientists did not understand the trypanosome life cycle well in 1900, but local Africans, European settlers, and colonial scientists all knew from personal experience that outbreaks of trypanosomiasis in humans and livestock were linked in some way to the simultaneous presence of wild animals and tsetse flies. This knowledge made many of them reluctant to set aside special areas as parks and reserves, when they might only promote the spread of the fly and the disease. Sensitive to this concern, colonial administrators convoked a series of tsetse fly conferences (in 1907, 1920, 1925, 1933, 1935, and 1936) and sought the advice of prominent scientists, including David Bruce, Charles Francis Massy Swynnerton, Robert Koch, and R. H. T. P. Harris. Unfortunately, these efforts resulted only in contradictory opinions and a hodgepodge of policies. In some areas, the tried-and-true practice of indigenous Africans—burning undergrowth and thicket, the favored habitat of the tsetse—was successfully employed. More often, as in the Tanganyika Territory, human populations were forced to move out of tsetse-infested regions and were resettled elsewhere, following the principle of human-animal segregation. Some governments had success with the Harris fly trap, which (as the name implies) reduced tsetse fly numbers by luring them into traps, but this was a labor-intensive and costly approach to tsetse control. More far-fetched was the British-German Treaty on the Combat of Sleeping Sickness in East Africa (1908), which declared war on crocodiles and crocodile eggs on the grounds that they were the main vectors of disease transmission. All of these efforts, whether effective or not, had the same basic goal: to intercept at some point the three-way link between wild animals, the tsetse fly, and human settlements.[73]

Many proposed a more draconian solution—the complete eradication of game from infected areas. "My honest conviction is that the presence and increase of game is entirely responsible for the presence and increase of tsetse, and that our game regulations are mainly, if not wholly, responsible for the increase of game," Rev. George Prentice wrote to the acting governor of Nyasaland in 1910:

> I hold that those who are responsible for the game laws are responsible for the presence of tsetse, and that the victims of trypanosomiasis are martyrs to the foolish policy of game protection. Any official, high or low, or any member of the Society for the Preservation of the Fauna of the Empire, who, in the face of known facts, asserts the contrary may prove the sincerity of his assertion by allowing us to experiment upon him with our local forms of tsetse. Until he does so, either his sincerity or his courage is open to question. But perhaps nothing is to be gained by going over past policy. What concerns us is the future and the present. There is a danger that from former statements that "there's no increase of tsetse," "there's no increase of game," and other equally stupid and childish assertions, we move to the opposite extreme and say "the infested area is too extensive," "the sacrifice of game would be too great." *No matter what the size of the tsetse-infested area, it must be tackled now. No matter what the sacrifice of game, it must be made now.* Slackness, delay, vacillation in the past have already produced disastrous results. Further delay would be criminal.[74]

The eminent British entomologist David Bruce concurred. "My advice is to clear out the game," he told the Interdepartmental Committee on Sleeping Sickness (a British investigatory team) in 1914, when asked what policy he thought the Colonial Office should follow in tsetse regions: "It would be quite as reasonable to allow mad dogs to run about English villages and towns under the protection of the law as to allow this poisonous big game to run about in the fly country of Nyasaland."[75] Following this advice, the governments of Southern Rhodesia and Natal (two regions where resentment toward the parks and nature reserves was high anyway) undertook massive animal-eradication campaigns over the next several decades, which resulted in the slaughter of perhaps three-quarters of a million wild animals.[76]

The tsetse fly question put the Fauna Society on the defensive almost from its creation. Although game-eradication policies did not violate the letter of the 1900 convention (Article III permitted eradication programs as long as they were "desirable for important administrative reasons"), they violated the spirit of the convention and called into question the appropriateness of game reserves and national parks. Edward North Buxton, the Fauna Society's first president, acknowledged to the Colonial Office in 1905 that "the tsetse fly disappears" in regions "where game has been totally destroyed," but he pointed out that "the danger of tsetse fly" was being invoked to justify the "wholesale destruction of game" even in areas where there was no problem:

> Now I do not think that is fair; it is as if you took a dozen men, one of whom you know had committed a crime, and put them all in prison. Who knows which species of animals are the "hosts" of the bacillus which is carried by the tsetse fly? It seems to me unjust that you should bring them all in guilty before you know, and kill them all because some of them may harbour the bacillus. Science has not yet arrived at the point that you can justly condemn all species; and we deprecate its being used as an excuse for the destruction of game generally—because all the species are held, without proper investigation, to be responsible for the continuance of horse-sickness."[77]

A few years later, Buxton again wrote to the Colonial Office:

> The game is spread over the country, but the fly—Glossina morsitans—is confined to very limited areas. It is not the case that the fly follows the game in their migrations, except for very short distances. These observers tell me that there is no general and obvious connection between the various species of big game and the fly, except that the latter are blood-suckers. The fly has been found plentiful in districts where the observers have seen no game, and there are large areas where game is abundant and there is no fly. The question which, as it seems to us, remains to be proved is by what species the trypanosome of nagana is really nurtured—if it can maintain existence in the blood of all, or only a few, or only one? If by all warm-blooded creatures, there is no proof that even the destruction of big game would meet the case. The infinitely more numerous small mammals, reptiles, and birds might continue to serve as the hosts of the trypanosome, and the larger animals might have been extinguished in vain.[78]

The Fauna Society had to deal with other park-related problems as well. An ideal game reserve, as defined by the 1900 London Convention, had to be large enough to incorporate the migratory patterns of the herds and have sufficient water holes, salt, and food for the migrating herds. The rough-and-tumble of colonial affairs, however, made it impossible even to remotely approach this ideal anywhere in Africa and most especially in the southern region. President Paul Kruger of the Transvaal, for instance, declared the area around the Pongola River a game reserve in the 1890s because it was malaria-infected and sparsely populated by whites *and* because it was situated along a disputed border with Great Britain (Kruger thought he could use the reserve as a bargaining chip in his negotiations with the British). Pongola was never rich in game in the first place, and the game that did roam there had been largely depleted by its game warden and by soldiers during the Anglo-Boer War. The reserve was even deproclaimed and reproclaimed several times before it finally ceased to exist in 1921.[79] The Sabi Reserve, also in Transvaal, had a far more fortunate fate: it became Kruger National Park in 1926, the second protected area in Africa to get the park designation. But for the first two decades of its existence, Sabi's first game warden (James Stevenson-Hamilton) spent most of his time doing battle with farmers who wanted to graze domestic cattle in its borders, with mining companies interested in the region's resources, with the builders of the Selati Railroad, and with soldiers (during the Anglo-Boer War and World War I) involved in guerrilla warfare.[80] Similar problems beset park wardens elsewhere.

The 1933 London Conference and the Apartheid Solution

In 1930, the Fauna Society sent R. W. G. Hingston to Africa on a fact-finding mission to determine how to address interrelated problems of ivory smuggling, animal cullings, and tsetse fly infestations. He concluded that the 1900 London Convention was functioning like bad "brakes" on the "destructive machinery" of colonial conservation and that, as a result, "African fauna is steadily failing before the forces of destruction brought to bear against it."[81] There were, he argued, four main causes of the destruction. The first cause was the spread of cultivation, which put farmers in increasing conflict with wild animals. "Man, once he cultivates an acre of soil, will not tolerate wild animals in his vicinity," he argued. The second cause was the trade in tusks, skins, and hides, which required the killing of the animals to obtain the products. The third cause was the hunting practices of indigenous Africans, who employed "methods that are wholesale and indiscriminating in their destructiveness." The fourth cause was the tsetse fly menace, which made so many colonists hostile to game protection.[82]

"How can this complex problem be dealt with in such a way as to lend some hope of preserving the species far into the future?" he asked rhetorically. "There would appear to be only one way. The human life and the wild life must be separated permanently and completely." As long as humans and animals were forced to live side by side, he argued, there would be demands to exterminate the local wildlife: "In one place the complaint will be that the crops are ruined, in another that the wild life kills domesticated stock, in another that it terrorizes the district, in another that it spreads disease." He concluded that the only solution to these problems was to separate humans and animals into "two completely distinct compartments." For animals to survive in modern Africa, he declared, they "must be segregated in a sanctuary."[83]

Hingston proposed the immediate establishment of a network of permanent nature parks large enough to offer genuine long-term protection to the whole gamut of the continent's animals. He noted:

> The weak point about the reserve is its insecurity and want of permanency. It is brought into existence by a Proclamation in the local Government *Gazette,* provided that the Secretary of State agrees. It can be removed by the same easy means. Should at any time a demand arise for a portion or the whole of a game reserve to be allocated to some other purpose, as for instance, agricultural development, it is not easy for even the Home Government to resist the demand and in practice it is not always resisted. In point of fact the game reserves of Africa are from time to time contracted, abolished, or altered in some way by this type of legislation. It is only a matter of time before a public demand will arise for the reserves or some portion of them to be thrown open, and there is no guarantee that any game reserve in Africa will last over an extended period.[84]

A policy of animal segregation, he argued further, would help wean Africans from the "primitiveness" of subsistence hunting and thus allow Europeans to teach them "the meat-securing methods which are practiced by more cultured races," namely, the "keeping and breeding of domestic animals such as cattle, pigs, goats, sheep, fowls and ducks."[85]

Nudged by the Fauna Society, the British government asked its Economic Advisory Council in 1931 to explore the possibility of a new international accord that would focus on making nature parks a permanent part of the African landscape. At Britain's urging, the International Congress for the Protection of Nature, which held its third (and final) meeting in Paris in July 1931, endorsed a revision of the 1900 London Convention. Then, in 1932, the British government established the Preparatory Committee for the International Conference for the Protection of the Fauna and Flora of Africa. It consisted of representatives from the Foreign, Colonial, and Dominion offices; the Fauna Society; the British Natural History Museum; Kew Gardens; the London Zoological Society; and the Economic Advisory Council, under the chairmanship of the Earl of Onslow. Its Draft Second Report served as the basis for the Convention Relative to the Preservation of Flora and Fauna in Their Natural State in 1933 (hereafter the 1933 London Convention).[86]

The Draft Second Report relied heavily on Hingston's analysis, though the authors put less blame on indigenous black populations and more on the colonial settlements for causing most of the disruptions. "The danger to any species of wild animal arising out of indiscriminate killing for sport or profit needs no emphasis," the introduction noted. "The increase of population also and the spread of cultivation and settlement, assisted by modern methods of irrigation and modern sanitary and medical knowledge, must lead in time to the disappearance of wild life from many areas in which it is now found." The committee saw two interlocking dangers to the viability of wildlife populations—"on the one hand the destruction of animals by hunters, often for commercial purposes, on the other the advance of settlement and the gradually changing character of the country." Agricultural and industrial developments in Africa were proceeding at a slow but steady pace, the report noted, and eventually, their combined impact would be felt throughout the continent: "In urging the need of protection of the wild life in Africa, we are not advocating a policy which is in any way inconsistent with the future destiny of the country. We call rather for the exercise of prudence and foresight in the conservation of an important part of its natural resources."[87]

The Draft Second Report emphasized that the primary purpose of the 1933 London Convention ought to be the "concentration of fauna in specially constituted sanctuaries." Much of Africa consisted of thinly settled regions where the local populace depended on agriculture and stock raising and where wild animals were often perceived as a nuisance, the report noted: "The harm done by marauding elephants and other animals to crops in many areas is only too evident. Indeed, in some British territories, the existing Game Departments had their origin in organizations the object of which was primarily the protection of crops of the natives from damage done by elephants and other wild animals." Domestic animals were also subject to diseases such as rinderpest and trypanosomiasis, wherein the "proximity of wild animals" to the domestic herds often accelerated the infection rates. "In many parts of Africa," the report added, "there is no graver problem affecting human welfare than the tsetse problem. Large areas of country which might be put to profitable use for grazing have had to be abandoned owing to tsetse infestation." Echoing Hingston, the report called for a system based on human-animal apartheid: "A final solution of the difficulties which arise from the intermingling of wild animals with native settlements can only be provided by the establishment of permanent and semi-permanent sanctuaries in which the animals can be effectually segregated."[88]

Preparations for the 1933 conference were so thorough that little discussion occurred at the meeting itself, and the plenipotentiaries of the Union of South Africa, Belgium, the United Kingdom, Egypt, Spain, France, Italy, Portugal, and the Anglo-Egyptian Sudan signed a final text that was nearly identical to the recommendations enumerated in the Draft Second Report. The prologue reiterated the principal goals of the 1900 London Convention but prioritized them differently. The main goal now was the establishment of "national parks, strict natural reserves, and other reserves within which the hunting, killing or capturing of fauna, and the collection or destruction of flora shall be limited or prohibited." Relegated to second place was the "institution of regulations concerning the hunting, killing and capturing of fauna outside such areas" and the "regulation of the traffic in trophies." Lowest on the list of priorities was the "prohibition of certain methods of and weapons for the hunting, killing and capturing of fauna."[89]

Article 1 declared that the convention would cover "all the territories (that is, metropolitan territories, colonies, overseas territories, or territories under suzerainty, protection or mandate) of any Contracting Government which are situated in the continent of Africa, including Madagascar and Zanzibar," and "any other territory in respect of which a Contracting Government shall have assumed all the obligations of the present Convention."[90] This made its geographic reach much more extensive than that of the 1900 treaty, which had not covered the territory north of the twentieth parallel or the large islands off the east coast of Africa.

Article 2 spelled out in detail what was meant by the terms *national park* and *strict natural reserve*. It defined a national park as an area "placed under public control" by a competent legislature, so long as it was set aside for the "propagation, protection and preservation of wild animal life and wild vegetation" or "for the preservation of objects of esthetic, geological, prehistoric, archaeological, or other scientific interest for the benefit, advantage, and enjoyment of the general public." A strict natural reserve, by contrast, was any area where hunting, fishing, forestry, agriculture, mining, and drilling were forbidden, as were any activities that in any way disturbed the flora and fauna within the confines of the protected area. Although there were some possibilities for overlap in this terminology, nature tourism was generally perceived as the defining feature of a park, whereas habitat and species protection was the defining feature of a strict natural reserve. No hunting was permitted in either area, except as authorized by the presiding authorities (game wardens, colonial governments, and so forth) for purposes of culling or animal control.[91]

Articles 3 through 7 obligated the participating governments to establish parks or strict natural reserves within two years of the treaty's ratification. To accomplish this task, the governments were supposed to "control" (though not necessarily exclude) all "white and native settlements in national parks" so as to reduce the possibility of damaging the natural fauna and flora. They were also encouraged to establish "intermediate zones" around the parks and reserves in which the "hunting, killing and capturing of animals may take place under the control of the authorities of the park or reserve." They were further urged to choose sites "sufficient in extent to cover, so far as possible, the migrations of the fauna preserved therein" and also to preserve a "sufficient degree of forest country." Finally, they were encouraged to work with neighboring countries in the establishment of transnational parks.[92]

Articles 8 through 10 and the annexes addressed the topic that had dominated the 1900 conference: hunting. Article 8 left much of the earlier hunting regimen intact, especially licensing requirements, but it spelled out in far greater detail the species that were to receive protection and divided them into two groups: Class A, which included animals whose protection was a matter of "special urgency and importance," and Class B, which included animals that could only be killed with a game license but whose preservation did not require "rigorous protection." The "vermin" category completely disappeared, a major advancement from the 1900 treaty. Article 8 was also slightly more favorable toward hunting by indigenous peoples: "No hunting or other rights already possessed by native chiefs or tribes or any other persons or bodies, by treaty, concession, or specific agreement, or by administrative permission . . . are to be considered as being in any way prejudiced by the provisions of the preceding paragraph."[93]

Article 9 broached a topic not handled in the earlier convention: the taking of "trophies," defined as "any animal, dead or alive, mentioned in the Annex to the Convention, or anything part of or produced from any such animal when dead, or the eggs, egg-shells, nests or plumage of any bird so mentioned." Importantly, it also declared that all "found" elephant and rhinoceros tusks (old tusks picked up from the ground rather than from freshly killed animals) belonged to the government and not to the individuals who found them. The delegates added this article because over the preceding three decades, many customs officials allowed hunters to transport freshly killed animals across borders under the pretext that they were "trophies" or "found tusks." Article 10 made it illegal for hunters to shoot from motor vehicles and aircraft or to use either to cause herds to stampede. It also reiterated the previous ban on the use of poison or explosives for killing fish and the use of nets, pits, snares, and poisoned weapons for hunting animals—yet another sign that traditional methods were still largely viewed as primitive and cruel.[94]

The conference was brief, lasting only from October 31 to November 8, and there were no topics that caused heated debate. Changes to the Draft Second Report were minimal, and all were designed to strengthen the treaty rather than water it down. At the request of Belgium, the concept of "strict natural preserve" was added to Article 2, which not only enhanced the preservationist thrust of the treaty but also provided an alternative to the Anglo-American notion that protected areas should pay for themselves through tourism. Article 7 was enhanced with four new sections (5 through 8). The first three granted extra protection to Africa's forested areas and indigenous tree species, and the fourth encouraged the "domestication of wild animals susceptible of economic utilization." Articles 9, 11, 12, and 19 were also slightly expanded, reworded, or altered. The only major task that fell to the conference participants was to compile the annex (which had not been prepared in advance) and determine which species required which level of protection. This task, too, proved uncontroversial.[95]

The 1933 London Convention officially went into force in January 1936, after being ratified by Egypt, the United Kingdom, Belgium, Sudan, the Union of South Africa, Portugal, and Anglo-Egyptian Sudan. At the time of ratification, however, the Belgium government inserted a "reservation" that diluted the effectiveness of the treaty: "Elephants shall not be considered in the Belgian Congo or in Ruanda-Urundi as being included among the animals mentioned in Class B, but shall be understood to be included in Class A (elephants each tusk of which does not weigh more than 5 kilogrammes)."[96] In less bureaucratic language, this meant that the Belgian-controlled regions would continue to outlaw trade in immature ivory (tusks under five kilograms) but would not accept the new restrictions on mature ivory (tusks over five kilograms).

For the rest of the 1930s, the British government tried to extend the terms of the African Convention to the Asian region. The Economic Advisory Council once again asked the Earl of Onslow to preside over the new Fauna and Flora of Asia Committee (later renamed the Committee for the Protection of the Fauna and Flora of Asia, Australia, and New Zealand), which was all but identical in representation to the earlier Preparatory Committee that had prepared the 1933 London Convention. Initially, a majority on the committee assumed the 1933 London Convention could simply be extended to include Asia with some minor adjustments in terminology, but after listening to the arguments for a new treaty, they decided to start from scratch. "Certainly the adoption by foreign Asiatic Governments of measures to prevent smuggling from Africa of trophies (notably rhinoceros horns) is essential to the effective working of the African Convention," Simon Harcourt-Smith, the most outspoken advocate of a new treaty, argued: "Nevertheless in such countries as Siam and to a very much greater degree French Indo-China there is a real and pressing need for internal legislation if certain rare species of fauna are to be preserved from extinction, and I venture to suggest that no effective action will be taken by either of the Governments concerned if they are merely invited to accede to the whole or part of the Africa Convention."[97] Unfortunately, the committee spent the next several years composing a new text and trying to settle on the proper geographic boundaries for the new treaty, and before the Conference for the Protection of the Fauna and Flora of Africa and Asia could commence as planned on November 7, 1939, the outbreak of World War II forced its abrupt cancellation.[98] Plans to hold the conference after the war never materialized, in no small part because the colonial powers found themselves on the losing side of national independence movements in both Asia and Africa.

In 1652, when the Dutch first established a toehold on the Cape of Good Hope, lions and elephants roamed free, and Europeans found themselves largely confined to isolated ports along the African coastline. Three hundred years later, Europeans moved freely throughout the continent, whereas wild animals found themselves increasingly contained within the boundaries of nature parks and game reserves. This massive transformation occurred almost entirely during the half century that separated the Berlin Conference and the 1933 London Convention.

Apartheid was, in many ways, the logical outcome of Europe's political and economic priorities. Wherever the Europeans established themselves in Africa—in the southern regions first and then elsewhere—they simultaneously exploited the animal resources around them and carved out tracts of land for cultivation and pasture. These dual endeavors could not be sustained forever, for they led to both a steep decline in animal numbers and an ever-quickening reduction in animal habitat. For the first two centuries, the damage remained confined to a handful of regions, but the technological-scientific revolution of the mid-nineteenth century spread the disruptions throughout much of sub-Saharan Africa. Elephants, rhinos, hippos, and many other large mammals were now easier to kill, thanks to a new generation of high-powered rifles and accurate scopes. Railroads opened up previously isolated areas for exploitation, settlement, and cultivation. The demand for tusks, skins, feathers, eggs, and many other animal products stimulated a commodities trade that reached around the globe, both for "legitimate" (government-sanctioned) and "illegitimate" (fenced and smuggled) products. The thirst for gold, diamonds, rubber, coffee, and bananas played a role as well. As Europeans became more aware of Africa's natural resources and as they extracted these resources from the continent as if there were no tomorrow, they increasingly disrupted the ecosystems that had maintained a vast array of animals for thousands of years.

Racism too played a role in the apartheid solution. The colonists took European culture and values with them to Africa, and they judged African societies largely on the basis of how closely they approximated (or were willing to adopt) those same standards. Europeans had long ago eradicated or confined so-called vermin in their own countries in order to make room for cultivation and cities. They had enacted game laws and created hunting preserves and protected areas in order to maximize the annual "game crop." They had hired wardens to catch and punish poachers. They had, insofar as possible, isolated their towns and villages from wild animals. The idea that societies could (or even should) strive to coexist with animals ran counter to the sensibilities they took with them to Africa, as well as to their political and economic interests in Africa. Had they looked at the world differently, they might well have carved out vast territories for indigenous Africans and indigenous animals that preserved landscapes from the European impact instead of creating nature parks and game reserves that catered mostly to white hunters and tourists. The establishment of megazoos was a peculiar solution to a specific problem that could have been solved differently only if the colonists had been of a different mind-set.

The haphazard way that Africa was carved up between 1885 and 1900 was significant as well. A hegemonic power on the continent might have been more willing to set aside for special protection larger chunks of territory in a greater variety of settings, perhaps even leaving intact territories that belonged to some of the larger and more powerful indigenous groups. A hegemonic power certainly would have been in a better position to create uniform game laws, control the flow of trade, and suppress smuggling. But Africa was carved up in the same way that Europe had been sliced and resliced over the centuries: by war, diplomacy, and sheer happenstance. The homespun rivalries of the British, French, German, Italian, Portuguese, and Spanish spilled over to the colonial arena, sometimes more virulently than in Europe, but so did a proclivity to cooperate and compromise with neighboring powers. The number of agreements these powers reached during this period—on everything from free trade to nature conservation—is a testimony to their ability to promote their own interests through collective means. They shared an interest in suppressing indigenous hunting traditions and practices. They shared an interest in making sure African blacks did not have access to modern weaponry. And they shared an interest in maintaining a plentiful supply of game animals. It was only when individual greed far outstripped common restraint—as was almost always the case with ivory—that the cooperative impulse seriously faltered.

The apartheid solution was not uppermost in the minds of Hermann von Wissmann, Lord Salisbury, and the dozens of other statesmen, big-game hunters, and scientists who dreamed up the 1900 London Convention. Their experience with game laws and nature reserves in Europe simply did not prepare them for the problems they would face in Africa. What worked in Brandenburg-Prussia (Wissmann's birthplace) or Hertfordshire (Lord Salisbury's birthplace) did not necessarily work well in Africa: European game wardens had to deal with plenty of poachers but not with trypanosomiasis, elephant rampages, ivory poachers, skin dealers, man-killing lions, and colonial rebellions. The need for the 1933 London Convention became increasingly apparent over time, as the colonial governments grappled with the implications of their hunting and conservation policies. Conservationists were surprised at the ferocity of the resistance to the new hunting regimen, and they feared for the long-term prospects of animal protection in light of this resistance. Strict separation between people and animals seemed like a logical long-range solution.

For all its problems, the 1933 London Convention did much to conserve Africa's wildlife in the face of relentless development and demographic growth. The British version of a protected area—a nature park that sustains itself economically on tourism—has proven quite successful in the former East African colonies of Uganda, Kenya, and Tanzania, where Serengeti, Kilimanjaro, and many other parks draw millions of tourists to visit each year, though the artificiality of these entities is hard to miss. The Belgium version of a strict natural reserve—in which tourism is kept to a minimum or prohibited entirely—has been more problematic. On the positive side, it offers a more "natural" setting for the protected animals, but on the negative side, it offers fewer opportunities for revenue production and therefore fewer incentives to hire game wardens, leaving the regions more vulnerable to poachers and smugglers.

There was much speculation as to whether the parks and reserves would survive the African decolonization process in the 1960s and 1970s. Concerns began to subside in 1964 when Julius Nyerere, soon to be Tanzania's first president, issued the Arusha Manifesto, in which he pledged to uphold the integrity of the park system and to promote nature conservation in postcolonial Africa. Fears were further laid to rest when major African leaders met in Algiers in 1968 and signed the African Convention on the Conservation of Nature and Natural Resources, which largely reaffirmed the 1933 London Convention. By the late 1960s, Africa's nature parks and reserves had become so famous around the world—and such a valuable source of tourist revenues—that it seemed folly to destroy them.

Time Line of African Animal Protection

1884   Chancellor Otto von Bismarck of Germany convoked Berlin Conference, attended by fourteen European countries. The conference addressed Belgium's claim to the Congo basin and laid down the rules for further African colonization, setting off the "scramble for Africa" that lasted until 1900.

1889   King Leopold II of Belgium convoked the Brussels Conference, which restricted the type of firearms and ammunition that could be sold to black Africans between the twentieth parallel north and twentieth parallel south and sanctioned the introduction of colonial gun licenses and big-game hunting restrictions.

1892   The Congo Basin Convention was signed by the Congo Free State (King Leopold II's personal fiefdom), France, and Portugal in Brussels to regulate export duties on elephant tusks in Central Africa.

1896   Hermann von Wissmann promulgated the German East African Game Ordinance, which served as a model for colonial game laws elsewhere in Africa and also as the basis for international agreements.

1900   The Convention for the Preservation of Wild Animals, Birds, and Fish in Africa (the 1900 London Convention) was signed in London in May 1900. Though never ratified, it led to greater uniformity in the regulation of Africa's migratory animals.

1902   The Society for the Preservation of the Wild Fauna of the Empire was established in Great Britain.

1909   The First International Congress for the Protection of Nature was held in Paris.

1923   The Second International Congress for the Protection of Nature was held in Paris.

1931   The Third (and final) International Congress for the Protection of Nature was held in Paris.

1933   The Convention Relative to the Preservation of Flora and Fauna in Their Natural State (the 1933 London Convention) was held in London in November 1933.

1934   The International Office for the Protection of Nature was established. It did not survive the impact of World War II.

1948   The International Union for the Protection of Nature, headquartered in Morges, Switzerland, was created. It became the International Union for the Conservation of Nature and Natural Resources in 1956 and the World Conservation Union in 1991.

1967   The Arusha Manifesto was declared by Julius Nyerere, who would soon become the first president of Tanzania. It established the framework for nature protection in postcolonial Africa.

1968   The African Convention on the Conservation of Nature and Natural Resources was signed in Algiers. It reiterated and expanded the terms of the 1933 London Convention.

1973   The Convention on International Trade in Endangered Species (CITES) was signed in Washington, D.C. It helped to halt the trade in wild animal products, including elephant tusks and rhino horns.

1979   The Convention on the Conservation of Migratory Species of Wild Animals was signed in Bonn (the Bonn Convention). It became first treaty of importance to protect migratory animals worldwide.

Chapter 2

The North American Bird War

> When I hear of the destruction of a species I feel just as if
> all the works of some great writer had perished, as if
> we had lost all instead of only part of Polybius or Livy.
>
> —Theodore Roosevelt (1904)

WILLIAM T. Hornaday's widely acclaimed book, *Our Vanishing Wild Life: Its Extermination and Preservation* (1913), reads like a war chronicle. As he wrote, Europe was poised for conflict over the Balkans, the United States was flexing its muscles in the Caribbean and Asia, and Mexico was in the throes of revolution. Hornaday, however, was neither a general nor a diplomat. He was the founding director of the New York Zoological Park (the Bronx Zoo) and one of the most avid conservationists of the Progressive Era. The war that prompted his book was the one being waged against the bird species of North America. "Throughout the length and breadth of America," he wrote, "the ruling passion is to kill as long as anything killable remains."[1]

As Hornaday saw it, this avian war was being fought by a vast "Army of Destruction" made up of six heavily armed divisions: "gentlemen sportsmen" (hunters who shot purely for pleasure), "game hogs" (trigger-happy hunters who always bagged their legal limit or beyond), "meat-gunners" (protein suppliers for big-city markets), "feather traders" (plume hunters for the hat-making industry), immigrant "slaughterers" (Italian Americans who used net traps, known as roccolos, to kill songbirds), and poor "southerners" (whites and blacks who lived off robins, mockingbirds, meadowlarks, and other "nongame" birds).[2] Based on the total number of hunting licenses that state governments issued in 1911, Hornaday reckoned the Army of Destruction was 2.6 million strong, not counting the "guerrilla army" (those who shot without licenses), whose numbers were unknown. "Indeed it is a motley array," he noted: "We see true sportsmen beside ordinary gunners, game-hogs and meat hunters . . . and well-gowned women and ladies' maids are jostled by half naked 'poor-white' and black-negro 'plume hunters.'"[3] Although Hornaday did not estimate the relative strength of each army division, he clearly feared the meat-gunners and feather traders most of all: they hunted for the market and not just for the pot.

Standing on the other side of the battlefield was a small "Army of the Defense," "friends of wild life who themselves are not on the firing line." This army consisted of federal officials, state game commissioners, bird conservationists, zoological societies, and recreational hunters. A veteran of earlier wildlife skirmishes (he was past president of the American Bison Society), Hornaday was pessimistic about the war's outcome: "Over the world at large, I think the active Destroyers outnumber the active Defenders of wild life at least in the ratio of 500 to 1; and the money available to the Destroyers is to the fund of the Defenders as 500 is to 1."[4]

One need not accept Hornaday's Manichean perspective or his racialized categories to appreciate the power of his metaphor. During much of the nineteenth century, Americans waged what can aptly be described as an unwitting—and ultimately self-defeating—war against their own wildlife. "There was a hazy kind of faith," John C. Phillips (one of the leaders of the Army of the Defense) later wrote, "in the existence far north of our borders of a sort of mysterious duck and snipe factory which could turn out the required supply practically forever."[5] This faith was misplaced. As railroads began to crisscross North America, as the American and Canadian populations moved westward, as farmers turned wetlands into fields, and as more and more commercial hunters blasted the sky for meat and feathers, the vast flocks of birds that once migrated across North America were becoming noticeably scarcer.

The front lines of the avian war ran north to south along the four great migratory routes of North America—routes Frederick C. Lincoln later named the Atlantic, Mississippi, Central, and Pacific flyways—that most bird populations used on their annual travels up and down the Western

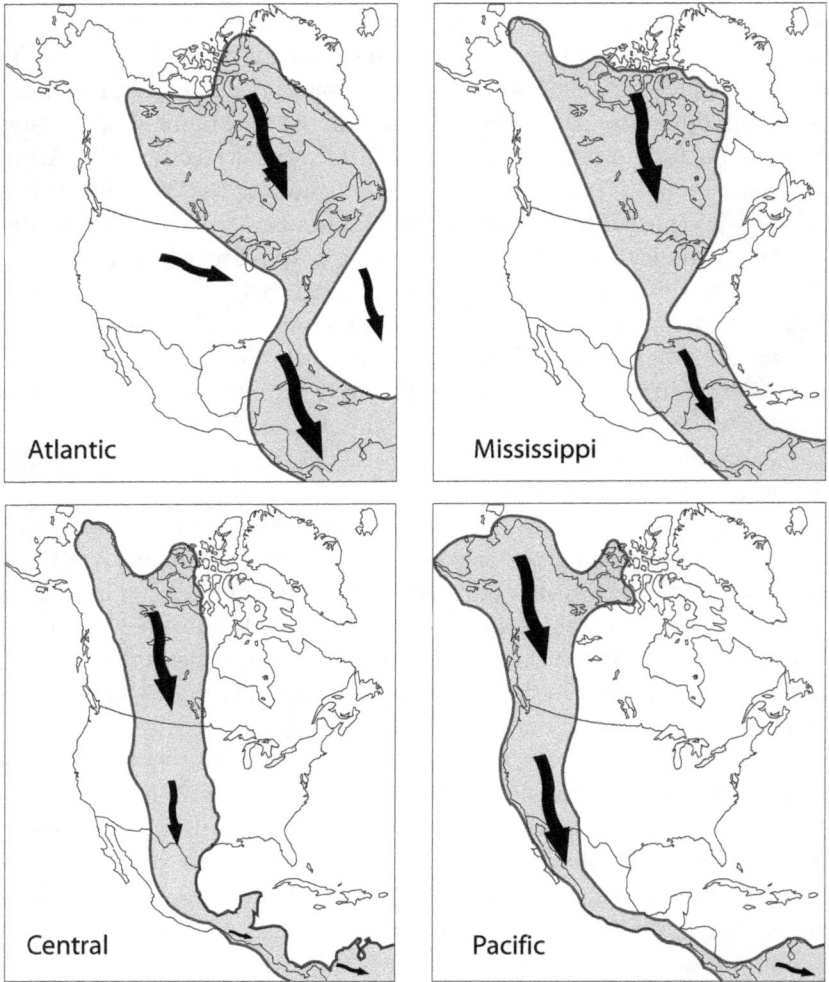

Figure 2.1. The four major flyways of North America: Atlantic, Mississippi, Central, and Pacific. Adapted from Frederick Lincoln, *The Migration of American Birds* (New York: Doubleday, 1939), 153, 167, 172, and 175.

Hemisphere. The main battlegrounds were the "staging posts" along these routes—places where birds collected to rest and feed before undertaking the next stage of their migration, many of which were located in the United States. By 1910, the nation's most renowned birding sites had become, as Hornaday noted, "so thoroughly 'shot out' that they have ceased to hold their former rank." Among these shot-out sites were Cape Cod (Massachusetts), the Great South Bay (New York), Currituck Sound (South Carolina), Marsh Island (Louisiana), the Sunk Lands (Arkansas), the Great Salt Lake (Utah), Klamath Lake (Oregon), the lakes of Minnesota, the whole Midwest, and all of southern California.[6]

The weapons of choice were the pump-gun, punt-gun, automatic rifle, and most especially the twelve-gauge shotgun, all weapons of capable of massive destruction. In 1911 alone, the country's four largest cartridge manufacturers (Winchester, Union Metallic, Peters, and Western) produced a combined total of 775 million shotgun cartridges, more than enough to pockmark the sky with buckshot. "It is natural," Hornaday noted, "for the duck-butchers of Currituck to love the automatic shot-guns as they do, because they kill the most ducks per flock. . . . It is natural for an awkward and blundering wing-shot to love the deadliest gun, in order that he may make as good a bag as an expert shot can make with a double-barreled gun. It is natural for the hunter who does not care a rap about the extermination of species to love the gun that will enable him to kill up to the bag limit, every time he takes to the field."[7]

The passenger pigeon was the best-known casualty of this warfare. Once by far the most plentiful bird in all of North America—with population numbers estimated at 3 to 5 billion—it was all but extinct by the beginning of the twentieth century (the last one dying in the Cincinnati Zoo in 1914), a victim of overhunting, habitat loss, and disease.[8] Nine other bird species—the great auk, the Labrador duck, the Pallas cormorant, the Eskimo curlew, the Cuban tricolored macaw, the Gosse's macaw, the yellow-winged green parrot, the purple Guadeloupe parakeet, and the Carolina parakeet—were also extinct or nearly so by 1910. Other birds—including the whooping crane, the trumpeter swan, the American flamingo, the long-billed curlew, the American egret, the snowy egret, the wood duck, the band-tailed pigeon, the heath hen, and the California condor—were directly in the crosshairs, their numbers so depleted at the time Hornaday wrote that extinction seemed only a matter of time. Shorebirds were especially endangered, among them the woodcock, the snipe, the willet, the dowitcher, the red-breasted sandpiper, the pectoral sandpiper, the upland plover, and the golden plover. Because market hunters could reach their rookeries with little effort, Hornaday predicted they would soon become "the first to be exterminated in North America *as a group.*"[9]

Gentlemen sportsmen, game hogs, meat-gunners, feather traders, immigrants, and poor southerners—as Hornaday well knew—were but the foot soldiers of this avian war. Two powerful warlords stood behind them, partially hidden amid the fog of battle: the meatpacking industry, congregated in Chicago, Boston, New Orleans, San Francisco, and a few other major cities, where millions of game birds were processed, sold, and shipped each year, and the millinery industry, headquartered in New York, which catered to the fashion in feathered hats. Behind these industries in turn stood American consumers, the real perpetrators of this war, and most especially urban middle- and upper-class whites, who found it sporting to dine on game birds and bedeck themselves with feathers. As historian Jennifer Price pithily noted, it was a time when "women were women, men were men, and birds were hats."[10]

Meatpackers were interested solely in "game birds," defined as any species for which there was enough consumer demand to make it worthwhile to kill, dress, package, and market them. Ducks, geese, swans, and other waterfowl were the mainstays, as were pigeons and doves. Culinary tastes for wild meat changed little over time, which meant that the same birds and the same staging posts were targeted year after year. There are no comprehensive records of the yearly kill, but anecdotal evidence (gleaned largely from law-enforcement records) suggests the numbers must have been in the hundreds of millions. During the winter of 1893–94, for instance, commercial hunters in Big Lake, Arkansas, shipped 120,000 mallards to market. At about the same time, some 120,000 robins were slaughtered annually in the cedar forests of central Tennessee.[11] Meanwhile, in 1902, investigators discovered a cold-storage facility in New York that contained over 40,000 illegally killed birds, including snow buntings, sandpipers, grouse, plovers, and quails. And in 1909, Louisiana reported a yearly kill of 5.7 million game birds (mostly wild ducks, quails, snipes, sandpipers, plovers, doves, geese, and brants).[12]

If meatpackers wanted game birds, the millineries sought "ornamental birds," defined as any species with colorful, large, or unusual plumage, the rarer the better. Most were not considered edible, so their carcasses were discarded after the feathers, wings, heads, and other fashion parts had been plucked and cut from them. Tanagers, cardinals, indigo buntings, orioles, blue jays, juncos, mourning doves, cuckoos, kingfishers, egrets, herons, owls, brown pelicans, and grebes were among the preferred species, but the fickleness of fashion ensured that the carnage spread across a wide range of birds all around the world.[13] The fall fashion of 1875 (the first year of the plumage craze) called for hats bedecked with hummingbird, pigeon, and bird of paradise feathers, as well as the wings of larks, blackbirds, and starlings. By the spring of 1884, stylish bonnets had to be adorned with the aigrettes (nuptial plumes) of herons or with marabou and hummingbird feathers. The 1896 winter season featured the wings, tails, and quills of grebes, parrots, and ostriches, and the summer vogue of 1899 brought forth walking hats covered with whole stuffed birds and golf hats ornamented with wings and feathers.[14] Though there are no reliable figures, conservationists estimated at the time that at least 200 million birds were sacrificed worldwide each year at the altar of hat vanity.[15]

Hornaday wrote in a period when the outcome of the avian war was still uncertain. In retrospect, it is easy to see that he greatly underestimated the size of the Army of the Defense and the weapons at its disposal. It consisted of three articulate and powerful divisions: bird conservationists (bird-watchers and bird scientists), recreational hunters, and farmers. Those in the first group—bird conservationists—came together in the 1880s under the aegis of two closely linked organizations: the American Ornithologists' Union (AOU) and the National Association of Audubon Societies (hereafter the Audubon Society). Aside from Hornaday, the most prominent leaders included George Bird Grinnell, editor of the sporting journal *Forest and Stream;* William Dutcher, first president of the Audubon Society and an AOU member; and T. Gilbert Pearson, second president of the Audubon Society and first president of the International Committee for Bird Preservation (established in 1922). The predominance of male names in the upper echelons of these organizations was deceptive, for almost all local and state bird chapters of the Audubon Society were established, led, and dominated by women—Harriet Hemenway, Mabel Osgood Wright, Minna B. Hall, Orinda Hornbrooke, and Jennie June Croly among them—even if national policies and diplomatic initiatives were largely in the hands of men.[16]

Recreational hunters made up the second group of bird protectionists. At the time, no clear line separated conservationists and sportsmen (Audubon, Grinnell, Dutcher, Hornaday, and Pearson were all penitent butchers, much like their counterparts in Africa). Nor was there a sharp break between those who hunted recreationally and those who earned some extra cash by supplying the local or national market; many hunters shot for food or sport on some days and for profit on others. Organizationally, however, recreational hunters constituted a distinct group: their political voice was the American Game Protective and Propagation Association (AGPPA), an organization inspired and funded by Remington, Winchester, Dupont, and other gun and ammunition manufacturers. Led by John Bird Burnham, AGPPA concerned itself exclusively with the conservation of *game* animals and the promulgation of laws to end *commercial* hunting. Some ornithologists shunned this organization (Pearson, for instance, declined the offer to become its first president), but many welcomed it as a useful counterbalance to the meatpacking lobby. Like many gun lobbies before and since, it also had plenty of money to throw around for purposes of propaganda and political influence.[17]

The third group of bird conservationists worked in the U.S. Department of Agriculture's Bureau of Biological Survey (hereafter the Biological Survey), which was later transferred to the Department of the Interior and renamed the Fish and Wildlife Service. Established at the urging of the AOU, the Biological Survey was entrusted with the task of protecting insectivorous and weed-eating birds—"birds useful to agriculture," in the language of the day—from decimation and extinction.[18] The Department of Agriculture estimated in 1904 that insect pests destroyed roughly 15 percent of the U.S. yield of cereals, hay, cotton, tobacco, sugars, fruits, and other farm and forest products each year, or $795 million out of a total of $5.5 billion.[19] In an era when chemical insecticides were still largely unknown, farmers and foresters were almost wholly dependent on birds to keep their crops and timber free of deadly pests. Unfortunately, many of these birds—including the rose-breasted grosbeak (which feeds on insect larvae), the scarlet tanager (tree lice), the cedar waxwing (cankerworm), the downy woodpecker (codling moth), the Baltimore oriole (cotton weevil), the upland plover (clover-leaf weevil), the killdeer (locusts), the sandpiper (grasshoppers), and the American goldfinch (weeds)—were being slaughtered en masse for their meat and feathers.[20]

The bob-tail quail was perhaps the most beloved—and endangered—of these agricultural birds, a fact that Hornaday used to great effect in his speeches and writings. As he wrote in *Our Vanishing Wild Life*:

> The next time you regale a good appetite with blue points, terrapin stew, filet of sole and saddle of mutton, touched up here and there with the high lights of rare old sherry, rich claret and dry monopole, pause as the dead quail is laid before you, on a funeral pyre of toast, and consider this: "Here lies the charred remains of the Farmer's Ally and Friend, poor Bob White. In life he devoured 145 different kinds of bad insects, and the seeds of 129 anathema weeds. For the smaller pests of the farm, he was the most marvelous engine of destruction that God ever put together of flesh and blood. He was good, beautiful and true; and his small life was blameless. And here he lies, dead; snatched away from his field of labor, and destroyed, in order that I may be tempted to dine three minutes longer, after I have already eaten to satiety."[21]

Prominent figures within the Biological Survey included George Lawyer, the first to promote a federal revenue stamp for waterfowl hunting (which later became the Duck Stamp); Theodore S. Palmer, who wrote the first draft of the U.S.-Canadian migratory bird treaty; Edward William Nelson, who helped compose the final treaty draft; and Jay "Ding" Darling, who was instrumental in the formulation of the U.S.-Mexican treaty and in the passage of the Duck Stamp Act. Although the Biological Survey's mandate extended only to insectivorous and weed-eating birds, its members understood that the preferred diet of this or that individual species was not a scientifically sound foundation upon which to base bird conservation, and from the outset, they worked closely with ornithologists to save a wide variety of birds and their habitats. There was, nonetheless, a limit to their largesse: birds that ate cereal crops (such as crows) and birds of prey (such as hawks and eagles) were not offered protection. As in Africa, these types of animals were considered vermin, for which the preferred remedy was eradication.[22]

In retrospect, it is easy to see that, in many ways, Hornaday misunderstood the nature of the war that he was chronicling. To begin with, the Army of Destruction was not as solid a force as he assumed. As the feather carnage became better known, as crops were repeatedly devoured by insects, and as game birds became ever scarcer, many bird slaughterers switched sides and joined the Army of the Defense. Here and there, the slaughter continued unabated, but more and more Americans came to realize that the killing frenzy had to come to an end before bird populations plummeted beyond the possibility of replenishment (the disappearance of the passenger pigeon served as a wake-up call for many). Furthermore, over time, a new front began to open up in this war: the legislatures and courts of North America. This phase of the war pitted states' rights and provincial rights advocates in the United States and Canada against federal authorities, nationalists against internationalists, and strict constructionists against judicial activists. As the battlefront moved from the flyways to the political arena, the Army of the Defense swelled with legislators and judges with a Progressive Era belief in the power of government and the importance of conservation.

Four simple but ingenious conservation measures brought the slaughter to an abrupt halt in the first half of the twentieth century. The first was a ban on hunting during the spring and summer months (known colloquially as the spring shooting season). This measure put an end to the killing of females and their young offspring during the mating and breeding months. The second was the outlawing of all *commercial* bird hunting.

This move saved game and nongame birds alike from mass slaughter at the hands of the meatpacking and millinery industries, while leaving recreational hunters free to shoot during the open season. The third measure entailed complete protection for certain endangered birds in order to restore their depleted populations. The fourth was the establishment of special bird reserves (sometimes also know by the quasi-religious term *sanctuaries*). These reserves provided breeding, feeding, and wintering sites for both migratory and nonmigratory species.

Ultimately, the real enemies of bird protection were those who opposed uniform game laws, who saw no value in bird reserves, who believed that state and provincial authorities could protect migratory animals without federal and international oversight, and who put too much credence in the invisible hand of laissez-faire economics. Success in the avian war depended almost entirely on an active federal government and a strong spirit of international cooperation—and therefore, on the defeat of states' righters, isolationists, and free traders.

From State and Provincial Game Laws to Federal Protection, 1871–1913

For most of the nineteenth century, game laws within the forty-eight states that make up the continental United States (some of which were still territories at the time) tended to be lax, ineffective, or nonexistent. As of 1870, only one state had imposed a complete ban on spring shooting of game birds. Another eleven had modest spring restrictions, and the remaining thirty-six states offered no restrictions whatsoever.[23] Regarding nongame birds, only sixteen states offered any protection, typically of a minimal nature, such as the outlawing of Sunday hunting (Minnesota and Iowa), proscriptions on killing certain insectivorous birds (Connecticut and Pennsylvania), protection for a few favored species (Massachusetts, Rhode Island, and Maine), or a ban on shooting "harmless" and small birds (New Jersey, Kentucky, and Michigan).[24]

Bird laws gradually grew more stringent over time, especially in states that had strong AOU or Audubon chapters or effective hunting organizations. A precedent-setting Supreme Court decision—*Geer v. Connecticut* (1896)—also acted as a spur to legislation, putting to rest the question of whether the state "owned" the wild animals within its borders or whether they belonged to those who owned the land upon which they were shot. The *Geer* case revolved around a hunter convicted of transporting game birds across state lines, in violation of Connecticut law. The hunter contended that a state had no right to regulate this type of interstate commerce. Writing on behalf of the Court majority, however, Justice Edward Douglas White upheld Connecticut's law and established the state-ownership doctrine in the process: "The right to preserve game flows from the undoubted existence in the state of a police power to that end."[25]

After the *Geer* decision, state legislatures passed a flurry of game laws with little fear of a constitutional challenge, and by 1913, legal protection for birds had improved remarkably. There was also a modest trend in the direction of uniformity among the states. All 48 states exercised some level of control over the hunting season, all required nonresidents to purchase a hunting license, all banned the export of certain game birds outside state borders, and all but one (North Carolina) banned the sale of certain game birds in state markets. Collectively, these restrictions placed some limits on interstate commerce in game birds, though the birds under protection varied too much from state to state for the restrictions to be truly effective. In addition, 43 states (excepting Arkansas, Kentucky, North Carolina, Rhode Island, and Virginia) had instituted daily bag limits, and thirty-nine states (excepting Arkansas, Maine, Maryland, Mississippi, North Carolina, South Carolina, Tennessee, Virginia, and West Virginia) required resident licenses. Meanwhile, forty-four states (excepting Arkansas, Mississippi, Nevada, Virginia) had a team of state game wardens to enforce the laws.[26]

Game laws in Canada's thirteen provinces and territories (some of which were not yet distinct entities) were similar to and often more stringent than those in the United States. By 1913, all imposed a closed season on the most highly prized game birds. All but the Northwest Territories imposed daily bag limits on hunters and banned the export of certain game birds outside the province in which they were killed. All but the Northwest Territories and Prince Edward Island also banned the sale of some game species on the public market. As in the United States, there was a general trend in the direction of greater uniformity in the laws.[27]

Protection for nongame birds also gradually improved, mostly because the Audubon Society successfully lobbied for passage of the AOU Model Law in many state and provincial legislatures. The AOU Model Law placed a total ban on the killing of most nongame birds and gave special protection to all birds with coveted plumes. As of 1913, 39 states (excepting Arizona, Idaho, Kansas, Maryland, Montana, Nevada, Nebraska, New Mexico, and Utah) had passed some version of the AOU law, as had many Canadian provinces, though enforcement (as with the game laws) was spotty at best.[28]

Despite the undeniable improvements in the direction of uniformity, the overall quality of bird protection continued to vary greatly from state to state and province to province. Some states and provinces—especially Florida, Arkansas, and the Northwest Territories—had weak laws and few (if any) game wardens. Other states—California, Massachusetts, and New York—had more stringent ones (New York was in a class of its own with passage of the 1910 Baynes Audubon Plumage Law, which outlawed all commerce in wild bird feathers in the state). Only eighteen states put a total ban on spring shooting. Some states had long open seasons, and neighboring states had short ones. Some states made an effort to protect breeding and feeding grounds, and others did not. Some states and provinces patrolled their borders diligently, and others winked as contraband crossed state and national lines. None of the Atlantic states or provinces offered any real protection to shorebirds, many of which were being decimated for their feathers. And not a single state or province placed a total ban on commercial hunting.[29]

In the absence of uniform state and provincial laws or effective enforcement mechanisms, it was child's play for meat-gunners to shoot birds in one state or province and fence the contraband in a neighboring one. And in the absence of international regulations, it was equally simple for milliners to evade the intent of the AOU Model Law by importing ornamental feathers from outside the United States. By the 1890s, therefore, state game commissioners, bird protectionists, and sportsmen began to clamor more and more for federal legislation.

Finding a legislative majority in the U.S. Congress in favor of national bird protection, however, was no simple task: the idea of a powerful and active federal government was still hotly contested, especially in the southern states, where the states' rights cause was strong. The various presidents of the period—William McKinley (1897–1901), Theodore Roosevelt (1901–9), William Howard Taft (1909–13), and Woodrow Wilson (1913–21)—could generally be counted on to favor federal bird legislation, with the possible exception of Taft. But it was anybody's guess as to how the U.S. Supreme Court might rule on the matter, since *Geer v. Connecticut* applied only to the states, not to the federal government.[30]

Given the political and legal constraints, federal legislation progressed slowly and cautiously. In 1900, Congress passed the Lacey Game and Wild Birds Preservation and Disposition Act (hereafter the Lacey Act), named after Representative John Lacey (R-IA), the wildlife enthusiast who sponsored the legislation. It prohibited the transport of wild animals, including birds and bird parts, across state lines if the person involved was in violation of state law when and where the animals were killed. Signed by

President McKinley, it gave the Department of Agriculture jurisdiction over the law's enforcement (though it was expected that the department would rely primarily on state game commissioners). The Lacey Act caused little concern to states' rights advocates because it bolstered already existing state laws without imposing uniform standards. Unfortunately, it was also a weak law, with no federal enforcement procedures and no funds for wardens. It did not take long for commercial hunters to realize they could flout the law with impunity.[31]

The next step came in 1904, when Representative George Shiras (I-PA), an avid wildlife photographer with ties to the Republican Party, introduced a bill in Congress "to protect the migrating game birds of the United States." Supported by Roosevelt, the measure bounced around Congress for nearly a decade before being amended to include birds useful to agriculture (the so-called dickey-bird clause) and eventually passed as the Migratory Bird Act of 1913. It is more commonly known as the Weeks-McLean Law, after Representative John Weeks (R-MA) and Senator George P. McLean (R-CT), who were instrumental in securing its passage. A mere page in length, the Weeks-McLean Law was one of the greatest milestones in U.S. bird-protection history. It targeted commercial hunting and the illegal transport of migratory birds, as well as placing proscriptions on the killing of insectivorous birds. The most contentious part of the legislation read:

> All wild geese, wild swans, brant, wild ducks, snipe, plover, woodcock, rail, wild pigeons and all other migratory game and insectivorous birds which in their northern and southern migrations pass through or do not remain permanently the entire year within the borders of any State or Territory, *shall hereafter be deemed to be within the custody and protection of the Government of the United States,* and shall not be destroyed or taken contrary to regulations hereinafter provided therefore.[32]

Empowered by the custody and protection clause, the Department of Agriculture imposed a nationwide ban on spring hunting and began to construct a set of regulatory guidelines for state governments to follow. The era of national protection had begun.

Meanwhile, President Wilson's new administration secured passage of the Underwood Tariff Act of 1913, which included a "feather proviso" (paragraph 347, formulated by Hornaday and Pearson) to curtail the international feather trade. It stated: "The importation of aigrettes, egret plumes or so-called osprey plumes, and the feathers, quills, heads, wings, tails, skins, or parts of skins, of wild birds, either raw or manufactured, and not for scientific or educational purposes is hereby prohibited."[33] The Underwood Tariff passed handily, though not without a debate that revealed the woeful ignorance of many senators and members of Congress on conservation matters. Thus, James A. Reed (D-MO), addressing the question of protecting egrets, stated: "I really honestly want to know why there should be any sympathy or sentiment about a long-legged, long-beaked, long-necked bird that lives in swamps, and eats tadpoles and fish and crawfish and things of that kind. Why should we worry ourselves into a frenzy because some lady adorns her hat with one of its feathers, which appears to be the only use it has?"[34]

The Underwood Tariff Act stood on a firm constitutional foundation—the interstate commerce clause. The Weeks-McLean Law, however, was vulnerable to both political and legal challenges. Its backers had slipped it into the Department of Agriculture's appropriation bill as a last-minute rider, leaving its opponents no real opportunity for discussion or debate. President Taft had then signed the bill on his last day in office, without realizing that the rider had been attached (he later claimed he would probably have vetoed it had he known). This stealth tactic merely ensured that the law's opponents would vigorously dispute its validity in the courts.[35]

In fact, the Weeks-McLean Law drew seventeen legal challenges in its first year alone and three more in the following year. As of 1916, the number of pending cases in federal and state courts had grown to twenty, with another thirty-three coming down the pike. There was, moreover, no uniformity to the judicial opinions that came out of these cases. Federal district courts in Oregon, California, South Dakota, Nebraska, Minnesota, and Michigan upheld the constitutionality of the law, but federal district courts in Arkansas and Kansas and the state supreme courts of Kansas and Maine did not.[36] Two of the federal cases in particular captured national attention: *United States v. Shauver* (1914) and *United States v. McCullagh* (1915). In the *Shauver* case, a federal judge in Arkansas threw out the conviction of a coot-shooter, agreeing with the defendant that the federal government had no jurisdiction over Arkansas's wild animals. In the *McCullagh* case, a federal judge in Kansas invoked the delegated-powers doctrine to toss a case involving a duck hunter: "If the act in question shall, on any ground, or for any reason, be upheld and enforced, it must surely follow [that] the many laws of the separate States of this Union must hereafter be held inoperative, for there can be no divided authority of the nation and the several States over the single subject matter in issue." Two state cases also received a high level of publicity. In *State v. Savage* (1915), the Kansas Supreme Court ruled that "the natural flight of wild fowl from one point to another does not constitute 'commerce,' unless that word be expanded beyond any significance heretofore given it."[37] Similarly, in *State v. Sawyer* (1915), the Maine Supreme Judicial Court ruled that neither the commerce clause nor the general-welfare clause applied to the issue of migratory birds.[38]

Of these cases, only *United States v. Shauver* reached the U.S. Supreme Court, and it was never fully adjudicated. At the first hearing, in October 1915, only six of the nine justices were present, and they were evenly split. The case was argued again in 1916, this time in front of all nine judges. By that point, however, Chief Justice White—the author of the *Geer v. Connecticut* decision and a passionate champion of federal bird protection—had become convinced that a majority of the justices would declare the Weeks-McLean Law unconstitutional. He therefore used his powers as chief justice to delay a verdict until Congress had a chance to find a way around this constitutional impasse.[39]

As it turned out, the easiest way around the impasse was to wrap the Weeks-McLean Law into an international treaty. Treaty making clearly fell within the prerogatives of the federal government, and a migratory bird treaty would therefore rest on firm constitutional footing.[40]

The 1916 Convention for the Protection of Migratory Birds

It is not entirely clear who first came up with the idea of circumventing the constitutionality issue via the 1916 Convention for the Protection of Migratory Birds (hereafter the 1916 Convention). There is some indirect evidence to suggest that it came out of a private conversation between a State Department lawyer (Fred K. Neilsen?) and a Supreme Court justice (Chief Justice White or Justice Oliver Wendell Holmes?), but there is no official record of this conversation.[41] What is clear is that Senator Elihu Root (R-NY) was the first to suggest a bird treaty in public. Root was a well-known internationalist who had helped initiate a new era in U.S.-Canadian relations while serving as Roosevelt's secretary of state. In January 1913, he introduced a Senate motion asking the president to pursue an accord with the "Governments of other North American countries."[42] When it stalled, Senator McLean introduced a near-identical motion proposing "the negotiation of a convention for the mutual protection and preservation of birds." It passed in April 1913.[43]

President Wilson responded to the Senate resolution by dashing off a quick note to his secretary of state, William Jennings Bryan, that stated:

"Personally, I should very much like to do this."[44] Initially, Wilson and Bryan had two broadly based international treaties in mind. The first was a U.S.-initiated treaty to protect migratory birds in the Western Hemisphere and the Pacific, which would include the territories of Canada, Mexico, South America, and Japan. The second was a British-initiated treaty to halt "the traffic in birds' plumage for millinery purposes" worldwide.[45] Both initiatives, however, would have required extensive diplomacy, by which time (it was feared) the Supreme Court would have voided the Weeks-McLean Law. Racing against the clock, the Wilson administration decided to reduce the territorial scope to Canada alone and to narrow the focus to migratory birds only.[46]

Canada's national government responded favorably to a bilateral treaty, though negotiations did not proceed quite as quickly as the Wilson administration had hoped. Unlike the United States, Canada lacked a well-organized Army of the Defense poised to champion the cause of bird protection. The country had many notable bird organizations—the Thomas McIlwraith Field Naturalists' Club of Ontario among them—but none comparable in size and stature to the AOU or Audubon Society. Nor was there much of a groundswell of public support for protection; in fact, some Canadians distrusted the treaty idea simply because it originated in the United States. Missing too were well-funded sportsmen's lobbies, though the North American Fish and Game Protective Association (a joint Canadian-U.S. organization) pushed for the treaty, as did the U.S.-based AGPPA. Both the Grand Trunk Railway and the Canadian Pacific Railway lent some support, mostly because their executives realized that wildlife attracted train tourists to the nation's national parks. Also backing the treaty was the tiny South Essex County Conservation Club, founded by Jack Miner, a pioneer in the establishment of bird reserves in Ontario, and Percy Taverner, a self-taught ornithologist who worked at the National Museum of Canada (now the Canadian Museum of Nature) in Ottawa.[47]

In the absence of a powerful and active bird lobby, the task of securing a treaty landed largely on the shoulders of middle-level civil servants, chiefly C. Gordon Hewitt, the Dominion consulting zoologist; James Harkin, the commissioner of Dominion parks, Canada's national park service; Maxwell Graham, chief of the Animal Division of the Dominion Parks Branch; and Martin Burrell, the minister of agriculture, a lukewarm supporter at best. Clifford Sifton and James White, chair and assistant chair of the Commission of Conservation (a short-lived, quasi-private institution created by the Ministry of Interior in 1909), also played a role in the discussions.

As in the United States, Dominion officials found themselves treading largely on terra incognita. The British North American Act of 1867, which governed Canada's Dominion status within the British Empire, placed the country's natural resources under the jurisdiction of the provinces. Though the 1867 act did not specifically list wildlife as a natural resource, federal authorities had traditionally treated it as one and had left hunting issues to provincial discretion, except in those regions that were directly under federal jurisdiction (such as the Northwest Territories). Before usurping a traditional provincial prerogative, federal authorities would normally take the matter to the provincial parliaments for a vote. This path, however, was bound to be slow and arduous and, in the end, probably also futile. They therefore chose the more expeditious route of asking the provincial cabinets (the premiers and their cabinet members) to agree to the treaty in principle and leave it up to Ottawa to negotiate the details. Though faster, this path had one disadvantage: it required the *unanimous* approval of the provincial cabinets, which meant that the treaty's provisions were hostage to the whims of each and every provincial cabinet.[48]

Fortunately, the majority of cabinets—including those of Quebec, Ontario, Prince Edward Island, New Brunswick, Alberta, Manitoba, and Saskatchewan—saw the benefits of uniform game laws and gave their consent with few complaints or delays. Among the others, the Yukon Territory was inadvertently overlooked and thus not consulted until after the treaty was signed. Newfoundland, which would not become a province until 1949, remained outside the treaty's scope. The Northwest Territories was already subject to national jurisdiction. Nova Scotia came on board after being promised a revision in the hunting season for shorebirds. British Columbia was the sole holdout, mostly because hunters there wanted to keep their traditional open season (five and a half months for ducks, six months for geese) and because some provincial officials feared (incorrectly, as it turned out) that the treaty would not apply to the nearby Territory of Alaska.[49] Eventually, the premier agreed not to thwart the negotiations as long as the treaty incorporated language giving British Columbia special privileges regarding the hunting of certain game birds. Once all the hurdles had been cleared, Canadian authorities issued an "Order in Council" on May 31, 1915, formally approving a treaty.[50]

The treaty text itself went through just one revision before reaching final form. Theodore Palmer of the Biological Survey wrote the initial draft (hereafter the Palmer Draft). For the most part, it was a compilation of the 1900 Lacey Act, the 1913 Weeks-McLean Law, and existing Department of Agriculture game regulations. Edward Nelson and C. Gordon Hewitt formulated the second draft (hereafter the Final Text) in light of changes demanded by Dominion authorities, the recalcitrant Canadian provinces, and several U.S. states.[51]

The Palmer Draft was so well crafted that Nelson and Hewitt only had to make about a dozen alterations altogether. Most were minor changes, such as the insertion of the word *migratory* before *game birds* to ensure greater textual clarity. Three changes were more substantive. The first affected the dates of the closed season. The second addressed insectivorous and weed-eating birds. The third exempted Native Americans from some of the terms of the treaty. Diplomats, not bird experts, initiated all three of these alterations, and they reflected the *political* exigencies of the negotiating process. Without exception, they were made in order to bring farmers, reluctant legislators, or some other key constituency on board or to address the special needs of indigenous hunters. Collectively, they reduced the conservationist goals of the Palmer Draft to a discernible degree—but not enough to erode the support of bird-protectionist groups.

The preamble (in both the Palmer Draft and the Final Text) was wholly lacking in the flowery and high-minded language that characterized so many other treaties of that era. "The United States of America and His Majesty the King of the United Kingdom of Great Britain," it stated in part, "being desirous of saving from indiscriminate slaughter and of insuring the preservation of such migratory birds as are either useful to man or are harmless, have resolved to adopt some uniform system of protection which shall effectively accomplish such objects."[52] The sole purpose of the treaty was to create a common set of game laws. Protection was justified on two purely practical grounds: birds eat crop pests and humans eat birds. Only those migratory birds that flew across the U.S.-Canadian border (roughly the forty-ninth parallel) were to receive protection.

Palmer's preamble made only one short reference to agricultural birds: "Many of these species are of great value in destroying noxious insects or as a source of food but are nevertheless in danger of extermination through lack of adequate protection during the nesting season or on their way to and from their breeding grounds." Both Nelson and Hewitt, however, realized that textual additions were needed to ensure that farmers would support the treaty. The Final Text was therefore strengthened to read: "Many of these species are of great value *as a source of food or in destroying insects which are injurious to forests and forage plants on the public domain, as well as to agricultural crops, in both the United States and Canada,* but are nevertheless in danger of extermination through lack of adequate protection during the nesting season or *while* on their way to and from their breeding grounds." In addition, a whole new section (Article VII) was added to ensure that the treaty did not interfere with farm-management policies: *"Permits to kill any of the above-named birds, which under extraordinary conditions may become seriously injurious to the agricultural interests in any particular community, may be issued by the proper authorities of the High Contracting Powers under suitable regulations prescribed therefor by them respectively, but such permits shall lapse, or may be cancelled at any time when in the opinion of the said authorities the particular exigency has passed and no birds killed under this article shall be shipped, sold or offered for sale."*[53]

Article I expanded the scope of the preamble somewhat in terms of both the birds deserving protection and the definition of *migratory*. Most important, it enumerated three categories of birds (not two, as suggested by the preamble). The first category was "migratory game birds." This set included the order Limicolae (shorebirds) and the families Anatidae (waterfowl), Gruidae (cranes), Rallidae (rails), and Columbidae (pigeons). The second category, "migratory insectivorous birds," included bobolinks, cuckoos, flycatchers, grosbeaks, meadowlarks, robins, waxwings, and many other species deemed useful to farmers and foresters. The third category was "other migratory nongame birds," a catchall phrase that allowed for the inclusion of the auk, the heron, the puffin, the tern, and other endangered birds. It was a credit to the diplomats that they did not make any changes whatsoever to Article I of the Palmer Draft, even though all three categories included some species whose migratory routes did not necessarily entail a border crossing and even though the third category included many birds with no hunting or agricultural value.[54]

Article II established the framework for a uniform hunting season in both countries. It limited the open season on game birds to a maximum of three and a half months per year, while prohibiting spring and summer shooting. The Final Text mandated a closed season for most game birds from March 10 to September 1. For shorebirds, it began February 1 and lasted until August 15. States and provinces were free to choose the opening and closing dates of the shooting season in their own territories, so long as those dates were within the time frame of September 1 and March 10 (August 15 and February 1 for shorebirds) and did not exceed three and a half months in length. A year-round closed season was declared for all migratory insectivorous birds and other migratory nongame birds. It was symptomatic of the game-law mentality that the negotiators couched what was obviously meant as a permanent and total ban on the shooting of these birds in the language of a "close season" that "shall continue throughout the year."[55]

In the same vein, Article III established a "continuous close season" for a period of ten years on a number of endangered game birds, including the band-tailed pigeon, crane (little brown, sandhill, and whooping), swan, and curlew. All shorebirds—except the plover (black-breasted and golden), Wilson's snipe, woodcock, and yellowleg (greater and lesser)—received the same protection. Article IV extended special protection to the wood duck and eider duck, both of which were so beloved by trigger-happy hunters that their survival was in doubt. This article gave states and provinces considerable leeway in determining how best to protect these ducks. They could declare a year-round closed season for five years or more, establish duck refuges, or enact "other regulations as may be deemed appropriate." Theoretically, they could even continue to allow an open season on wood duck and eider duck, as long as they took adequate countermeasures to ensure a rebound of these species within their territories.[56]

Articles II and III of the Final Text differed in significant ways from the Palmer Draft. Palmer sought a considerably longer closed season—to begin on February 1 and last until September 1, dates that spanned the breeding season for nearly all bird species. However, at the insistence of fifty-two U.S. members of Congress from a handful of states (principally from Louisiana, Mississippi, Illinois, and Missouri), Hewitt and Nelson felt compelled to reduce the mandated closed season by nearly six weeks, so that it began March 10 and ended September 1. Initially, the Biological Survey stiffly resisted the new dates, as did Hewitt, arguing (with much justification) that it left mothers and their young vulnerable during the earliest part of the breeding season. But they yielded to the political pressure once they received assurances that the three-and-a-half-month–maximum open season would remain intact.[57]

The demands of Nova Scotia and British Columbia account for the remaining changes in the dates of the hunting season. The Palmer Draft called for a closed season on migratory shorebirds (Limicolae) from February 1 to September 1. Nova Scotia negotiated a shortening of the closed season by about two weeks—February 1 to August 15—for hunters residing on the Atlantic coast north of Chesapeake Bay (roughly the New England states and the Maritime provinces). This change did not trouble the U.S. negotiators because the Department of Agriculture already granted these same exceptions to the New England states in its domestic regulations. British Columbia, meanwhile, refused to adhere to the mandatory ten-year closed season on cranes, swans, and curlews, all of which were protected by Article III. To accommodate this objection, the Final Text gave that province's parliament leeway in deciding whether these game birds deserved protection.[58]

The Palmer Draft did not address the issue of indigenous hunting rights at all, even though it was common practice in the Territory of Alaska, the Northwest Territories, and elsewhere to exempt "Indians" and "Eskimos" from some game provisions. To rectify this deficiency, the Canadian government secured a slight change to Article II. Paragraph 1 (which dealt with the issue of closed seasons) was amended to read: "Indians may take at any time scoters for food but not for sale." Paragraph 3 (which protected certain endangered birds) was reworded to read: *Eskimos and Indians may take at any season auks, auklets, guillemots, murres, and puffins, and their eggs, for food and their skins for clothing, but the birds and eggs so taken shall not be sold or offered for sale.*[59] Indigenous hunters, however, were not exempted from the spring-shooting ban on other birds, even though other treaties guaranteed them the right to year-round hunting (this was not an oversight but rather a deliberate attempt by Canadian authorities to use this treaty to override previous agreements).[60] They circumvented the treaty as best they could, and sensible game wardens winked at their transgressions, but it remained a major focal point of discontent until the Protocol Amending the 1916 Convention for the Protection of Migratory Birds finally eliminated these restrictions in 1995.[61]

Neither the Palmer Draft nor the Final Text addressed the international feather trade directly, since a separate treaty was being contemplated at the time. Two articles, however, put limits on the commerce in other bird products, using language borrowed from the Lacey Act of 1900. Article V prohibited the "taking of nests or eggs of migratory game or insectivorous or nongame birds" except for scientific purposes. Article VI prohibited the "shipment or export of migratory birds or their eggs from any State or Province, during the continuance of the close season in such State or Province." The same article also prohibited the "international traffic in any birds or eggs" that were "captured, killed, taken, or shipped at any time contrary to the laws of the State or Province in which the same were captured, killed, taken, or shipped." The latter prohibition, which appeared in the Palmer Draft, was initially removed on the grounds that it strayed too far from the purpose of the treaty, but it was restored in the Final Text.[62]

Canada's quasi-colonial status within the British Empire delayed the ratification process. Negotiators in Ottawa and Washington began work on the Final Text in February 1914. After they finished, it then had to go to the British Embassy and from there to the Foreign Office, Colonial Office, governor-general's office, and finally back to the British Embassy. (Even U.S. authorities were left in the dark for a long while as to its whereabouts, not least because the Order in Council lay misfiled for months in the British Embassy.) The outbreak of World War I in Europe in August 1914 no doubt slowed the process as well. At long last, the British government agreed to the treaty in February 1916; the U.S. secretary of state, Robert Lansing, and the British ambassador, Cecil Spring-Rice, formally signed the treaty in August 1916; and both governments ratified it in December 1916. Cecil Spring-Rice's apt description of the treaty—as an agreement "by which the protection accorded to migratory birds in the United States under the law of March 4th, 1913 [the Weeks-McLean Law], should be extended to the Dominion of Canada"—clearly belied its U.S. origins.[63]

The 1916 Convention required each government to pass an enabling law, known in Canada as the Migratory Birds Convention Act of 1917 (MBCA) and in the United States as the Migratory Bird Treaty Act of 1918 (MBTA). Aside from reiterating the treaty's stipulations, these laws consolidated past federal regulations, established game warden agencies, and enumerated a system of penalties and fines for poaching (typically $10 to $25 in the early years). In Canada, oversight was given to the governor-general-in-council and the Department of Interior. In the United States, authority resided first in the Department of Agriculture's Biological Survey; it was then transferred to the Department of Interior's Fish and Wildlife Service. Both enabling laws gave these bodies a tremendous amount of leeway to update federal regulations year by year without first having to secure legislative approval. In practice, this has meant that federal, state, and provincial game wardens have been able to modify federal game laws in accordance with changing conditions.[64]

States' rights advocates in the United States—especially those in Missouri and Kansas—immediately challenged the constitutionality of the 1916 Convention and MBTA, though the treaty-making clause in the U.S. Constitution now put them at a great disadvantage. The main test case was *Missouri v. Holland* (1919). It pitted the attorney general of Missouri, Frank McAllister, who went duck hunting during the federally mandated closed season, against Ray Holland, the federal game warden who arrested him. McAllister's challenge rested entirely on the primacy of the Tenth Amendment, which gave the states authority over matters not specifically delegated to the federal government. Holland's attorneys relied on the treaty-making authority of the federal government. By a vote of seven to two, the Supreme Court upheld the constitutionality of both the treaty and the federal legislation. Writing on behalf of the Court majority, Justice Oliver Wendell Holmes dismissed the notion that the Tenth Amendment trumped the federal government's treaty-making authority: "Wild birds are not in the possession of anyone, and possession is the beginning of ownership.

The whole foundation of the State's rights is the presence within their jurisdiction of birds that yesterday had not arrived, tomorrow may be in another State, and in a week a thousand miles away." Bird protection was a "national interest of very nearly the first magnitude," beyond the capability of the states to handle effectively: "The subject matter is only transitorily within the State and has no permanent habitat therein. But for the treaty and the statute there soon might be no birds for any powers to deal with. We see nothing in the Constitution that compels the Government to sit by while a food supply is cut off and the protectors of our forests and our crops are destroyed."[65]

Unable to thwart federal jurisdiction on constitutional grounds, treaty opponents attempted to exploit ambiguities in the term *migratory bird* to gut its efficacy. This time, the main test case was *United States v. Lumpkin* (1921), which began when a Georgia hunter was arrested and fined $25 for killing doves during the federally mandated closed season. The doves in question were mourning doves (also known as turtle doves). A songbird, the mourning dove had replaced the vanished passenger pigeon as a favorite of hunters in the southeastern part of the United States (and it remains to this day a mainstay of game hunters). The hunter in the *Lumpkin* case contested his fine on the grounds that doves were not genuinely "migratory" because some mourning doves stayed within the territory of the United States the entire year, never reaching as far north as Canada. A Georgia district judge, however, disagreed: "I think what the Treaty means to say is this: Our purpose is to deal with migratory birds, but we do not want it left up in the air; we don't want it subject to uncertainties that will inevitably arise, and differences of opinion that will exist in various localities; we don't want hunters or birds under uncertainties of that sort, but will proceed to examine and find out and agree as to what kind of birds we are talking about; and they mentioned doves." And later in the opinion, he stated: "I think that this Treaty plainly states that it is agreed that doves (which certainly must have included turtle or mourning doves if it included any) are migratory."[66]

In Canada, the constitutionality of the treaty and enabling law was never seriously in doubt, but some hunters tested the measures nonetheless. The key case, *King v. Clark* (1920), revolved around the arrest of a hunter caught illegally shooting fourteen geese on Prince Edward Island. He challenged the arrest on the grounds that the British North American Act of 1867 gave the provinces jurisdiction over wildlife. After winning in municipal court, he lost on appeal to the Supreme Court of Prince Edward Island, the presiding judge citing Oliver Wendell Holmes in rendering his

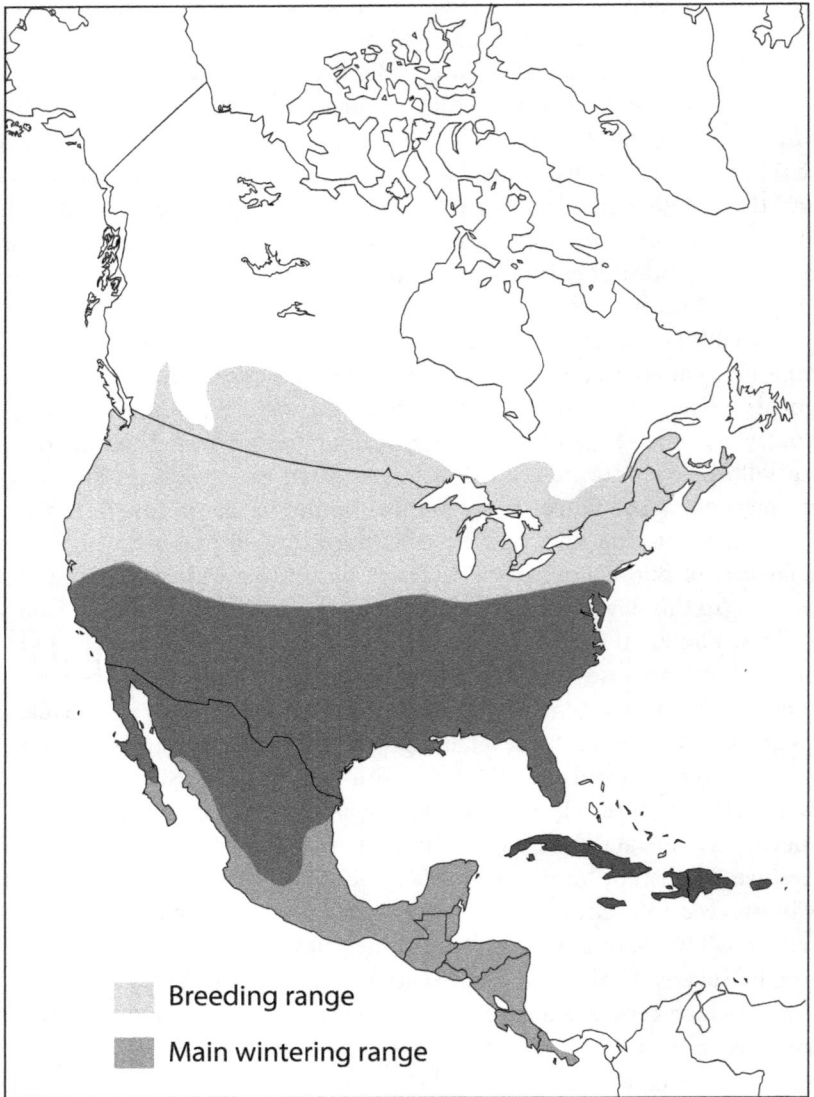

Figure 2.2. The breeding and wintering ranges of mourning doves (turtle doves). In *United States v. Lumpkin* (1922), a federal district judge determined that mourning doves were migratory even though some flocks did not cross the Canadian or Mexican borders during their annual movements. Thomas C. Tacha and Clait E. Braun, *Migratory Shore and Upland Game Bird Management in North America* (Washington, DC: U.S. Fish and Wildlife Service, 1994), 9.

verdict. When the Canadian Supreme Court declined to hear the hunter's appeal, his conviction stood. The constitutionality issue was settled.[67]

As in the United States, Dominion authorities had to deal not just with constitutional issues but also with some dissension among the provinces. Alberta, Saskatchewan, Manitoba, Ontario, and Quebec all quickly brought their provincial legislation into conformity with federal law as required by the treaty. So did the once-recalcitrant British Columbia, which chose not to take advantage of the exemptions that it had been granted. Dissatisfaction, however, was high among parliamentarians in the Maritime Provinces of New Brunswick, Nova Scotia, and Prince Edward Island, mostly because sportsmen there were reluctant to abandon their spring-shooting traditions, treaty or no treaty. To show its displeasure, the New Brunswick legislature repealed all of its waterfowl regulations, declaring that it was now up to federal authorities to impose and enforce game laws. Neither Nova Scotia nor Prince Edward Island legislators took such draconian steps, but they did let it be known that they would not underwrite the costs of hiring game wardens. "The expressions that I have heard," Prince Edward Island's premier stated at a meeting convoked to iron out federal-province differences, "have been to the effect that the legislation was designed for the benefit of the southern portion of North America." Eventually, Dominion authorities were forced to hire their own game wardens to patrol these provinces. The Yukon Territory, for its part, felt cheated by the fact that the fall hunting season did not begin until September 1, by which time most waterfowl and shorebirds had already headed south for the winter. This unfortunate circumstance (which affected the neighboring Territory of Alaska as well) could not be rectified without reopening negotiations, which neither Ottawa nor Washington was willing to do.[68]

Indigenous hunters fared no better. When the Fort Chipewyan chiefs in Alberta Province complained in July 1927 that the U.S.-Canadian agreement stood in crass contradiction to their prior treaty with the Canadian government, the Department of Indian Affairs simply told them that the Migratory Bird Convention Act trumped all prior commitments. "Are we then to starve during the summer months," the chiefs queried, "because the whiteman has broken his word to us?"[69]

The Path to Bird Reserves in the United States and Canada

American, Canadian, and British diplomats largely allowed bird conservationists and scientists to formulate the treaty's provisions, intervening only when it was necessary to guarantee a smooth ratification process. This approach ensured that the treaty had a much stronger conservationist edge than did the African treaties of 1900 and 1933, even if (as in Africa) it greatly shortchanged the indigenous populations. Altogether, nearly 540 species of migratory birds came within the scope of the treaty.[70] Game birds were protected during the breeding season, and insectivorous and endangered birds were protected year-round. Diplomats also allowed scientists to determine which birds should receive protection, even though the list included some that were clearly neither game nor insectivorous birds, as well as a few whose migratory routes were not fully known. As a rule, if conservationists believed that any particular migratory bird deserved protection, the diplomats went along with them, as long as some of the species within that family inhabited both the United States and Canada.

Although there was no explicit avenue for expanding or contracting this list after 1916, some of the treaty's terminology was ambiguous enough to allow for a degree of flexibility. Article I, for instance, listed birds by order or family, leaving it up to federal authorities to decide which species within these groups deserved protection. Similarly, the same article, after listing a few dozen insectivorous birds by species name, then turned around and granted protection to "all other perching birds which feed entirely or chiefly on insects"; this phraseology left plenty of room for expansion, since the phrase *perching birds*, if interpreted as being synonymous with the order Passeriformes, accounts for about half of all bird species worldwide, many of them insect-eaters. Over the next decades, both Canada and the United States did in fact utilize the textual flexibility to extend the treaty's protective net, until it protected nearly sixty families representing over seven hundred species.[71]

The 1916 Convention, however, had some serious flaws. It established a three-and-a-half-month-maximum hunting season, but it did not require any uniformity or cooperation among the various states and provinces (or even among the counties or districts within a state or province) in determining the exact dates. Staggered seasons were the outcome, potentially leaving some game birds "under fire" at one point or another on their migratory routes for nearly six months of the year (though in practice, it would be closer to four and a half months).[72] Furthermore, some 220 migratory species initially did not receive any protection whatsoever, either because they were not considered edible, because they did not feed primarily on insects and weeds, because they were birds of prey (and therefore deemed vermin), or because they were not yet endangered. Prominent among these nonprotected birds were the albatross, the cormorant, the pelican, the flamingo, the ibis, the hawk, the owl, the parrot, the kingfisher, the crow, the jay, the blackbird, and the mockingbird—some of which would become endangered in the decades after the treaty went into effect. Also specifically excluded from the terms of the treaty were all game birds considered resident (nonmigratory), including quails, pheasants, grouses, wild turkeys, and other gallinaceous birds beloved by hunters, some of which were already endangered.[73]

Nothing in the treaty, of course, prevented Canada and the United States or the individual states and provinces from enacting legislation on their own to protect species excluded from the 1916 Convention. But the initiative would have to come from below, which had predictable results: laws were slow in coming and poorly coordinated among neighboring territories. Before more-stringent measures were put in place, the heath hen would become extinct and the California condor nearly so.

By far, the most glaring weakness in the 1916 Convention was the absence of any paragraphs specifically devoted to the establishment of bird reserves. Only Article IV broached the subject of reserves at all and then only as an option for those states and provinces that chose not to institute a five-year closed season on endangered ducks. Those involved with negotiating the 1916 Convention certainly appreciated the importance of habitat protection, but they were far more preoccupied with what seemed at the time a more urgent concern—saving overexploited birds from the brink of extinction. The treaty, therefore, provided no grand design for saving important wetlands and staging posts. Bird reserves thus emerged in haphazard fashion, slowly over time.

President Theodore Roosevelt (largely on the advice of the AOU and the Audubon Society) had already begun to establish federal bird reserves in the United States, a full decade before the bird treaty was conceived. In keeping with the temper of the times, these reserves were not designed to provide a large and diverse habitat for all birds but simply to preserve a few endangered species at a handful of key locations. In practice, this approach largely meant protecting Atlantic shorebirds from the predations of meat, egg, and feather hunters. In 1903, Roosevelt issued an executive order declaring the tiny government-owned Pelican Island (Florida)—an endangered brown pelican rookery—as the first federal migratory bird refuge. He created another fifty-five reserves by executive order during his time in office, including the Klamath Lake Reserve (Oregon), the first designed specifically to protect waterfowl, and the Hawaiian Islands Reservation for Birds, the first in the new Territory of Hawaii. Subsequent presidents followed suit, and by 1929, the United States possessed over seventy federal refuges nationwide by way of executive order.[74] Some state legislatures also created special wildlife reserves, with California leading the way in 1870 with the establishment of Lake Merritt (now just a pond in the middle of downtown Oakland); California was followed by Indiana (1903), Pennsylvania (1905), Alabama (1907), Massachusetts (1908), Idaho (1909), and Louisiana (1911).[75]

Meanwhile, the Audubon Society set up a system of private refuges funded by its own membership and outside donors. By 1929, it had created fifty-nine sanctuaries in twelve American states, forty-two of which were along the Atlantic shoreline: Maine (ten), Connecticut (six), New York (two), New Jersey (one), Virginia (two), North Carolina (one), South Carolina (two), Georgia (three), and Florida (fifteen). Some of these refuges—such as the twelve-acre Roosevelt Bird Sanctuary surrounding the president's grave at Oyster Bay in Long Island, New York—were more symbolic than anything else. But others—such as the twenty-six-thousand-acre Paul J. Rainey Wildlife Sanctuary (Louisiana) on the Gulf of Mexico—were large enough to provide protection for a wide variety of small land birds, waterfowl, and shorebirds.[76]

Congress too moved slowly but inexorably in the direction of a federally funded reserve system. In 1922, Representative Daniel Anthony (R-KS) proposed a federal license of $1 per year applicable to hunters in all states. The money generated by the licenses was to be designated for the purchase of land for refuges and public shooting grounds, as well as for the salaries of game wardens. "The result of the migratory bird treaty act has been a great increase in the supply of wild fowl," testified Edward Nelson, chief of the Biological Survey and cowriter of the 1916 Convention, appearing before a House of Representatives subcommittee in defense of Anthony's bill, "but the rapid increase of drainage throughout the United States is taking away the homes of these birds, and if this is continued without any effort being made to maintain marsh and water areas for the birds the ultimate result will be the wiping out of the birds, simply because they will have no place in which to live, to breed, and to feed."[77]

The Anthony bill never became law, mostly because states' righters opposed the concept of a federal license and because some conservationists (notably Hornaday) disliked the idea of creating public shooting grounds.[78] Subsequent efforts, however, were more successful. In 1924, Congress authorized $1.5 million to purchase nonagricultural land along the Mississippi River between Rock Island, Illinois, and Wabash, Minnesota, for the establishment of the Upper Mississippi Wildlife and Fish Refuge, an important staging post on the Mississippi flyway. In 1928, it authorized funds for the creation of the Bear River Migratory Bird Refuge near the Great Salt

Figure 2.3. The Duck Stamp (federal wildfowl hunting license) for the 1934–35 season. Migratory Bird Conservation Commission, *2000 Annual Report: Migratory Bird Conservation Commission* (Washington, DC: Migratory Bird Conservation Commission, 2000), front cover.

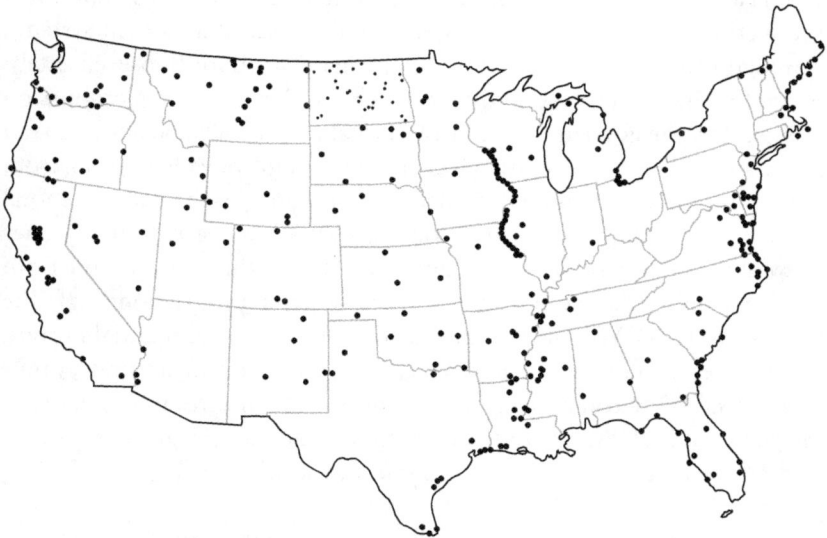

Figure 2.4. National Migratory Bird Refuge Areas in the continental United States. Migratory Bird Conservation Commission, *2000 Annual Report: Migratory Bird Conservation Commission* (Washington, DC: Migratory Bird Conservation Commission, 2000), 9.

Lake in Utah, and in 1930, it did so for the Cheyenne Bottoms Migratory Bird Refuge in Kansas; both refuges were along the Central flyway.[79]

Then, in 1929, Senator Peter Norbeck (R-SD) and Representative August Andresen (R-MN) managed to steer the Migratory Bird Conservation Act (or Norbeck-Andresen Act) through Congress, along with a $7.8 million appropriation for land purchases over the next ten years.[80] The act led to the establishment of the Migratory Bird Conservation Commission, composed of the secretaries of agriculture, commerce, and interior as well as two senators and two members of Congress, to oversee the purchase of land for bird reserves. In 1934, Congress took yet another step, passing the Migratory Bird Hunting Stamp Act (better known as the Duck Stamp Act). Its biggest champion was Ding Darling, a popular political cartoonist who served briefly as chief of the Biological Survey during the presidency of Franklin D. Roosevelt (1933–45). The act required hunters to purchase a $1 stamp each year (a disguised federal license) for the right to shoot wildfowl, with the proceeds earmarked for land purchases.[81]

These congressional actions added many elements that were missing in the 1916 Convention and MBTA, setting the stage for the establishment of today's National Wildlife Refuge System (formally constituted in 1966, when game ranges, wildlife ranges, wildlife-management areas, waterfowl-production areas, and wildlife refuges were consolidated under one system). The figures tell the tale. Between 1903, when Theodore Roosevelt established the first bird refuge, and 1929, when the Conservation Commission was created, the government set aside a total of only 668,000 acres of land and water for bird protection. Over the following twenty-five years, from 1929 to 1954, the government created 129 more refuges, bringing the total to nearly 3.8 million acres, more than a fivefold increase. Roughly one-third of the commission's land acquisitions during that period came from Duck Stamp funds, the rest from congressional appropriations.[82] By the time the National Wildlife Refuge System celebrated its centennial in 2003, the number of refuges had grown to nearly 540, covering around 95 million acres.[83] (One could plausibly add the 83 million acres in 369 parks of the National Park System to the tally; though not specifically designed as bird refuges, these parks do provide safe and secure breeding and feeding grounds.)

Aside from administering the refuges, the Biological Survey issued new regulations in the 1920s and 1930s designed to protect both bird habitats and bird species.[84] Hunting was temporarily banned year-round along certain popular stretches of the Mississippi flyway to give duck populations a chance to rebound. Bag limits were reduced to twenty-five per day for soras;

twenty per day for Wilson's snipes; fifteen per day for ducks, rails, and gallinules; and four per day for woodcocks and geese. A year-round closed season was established for black-bellied and golden plovers and greater and lesser yellowlegs throughout the country.[85] Most controversially, the Biological Survey banned the practice of baiting (sprinkling grain and other food around hunting sites to lure birds), and it restricted the establishment of private shooting grounds adjacent to federal reserves. In two key test cases, *United States v. Reese* (1939) and *Bailey v. Holland* (1942), the courts upheld the constitutionality of these regulations.

The establishment of a bird reserve system proceeded more smoothly in Canada than in the United States, though the total acreage under protection in the first decades was roughly the same (640,000 acres in Canada as of 1933, as compared to 668,000 in 1929 in the United States).[86] In 1887, Dominion authorities established a bird reserve at Last Mountain Lake in Saskatchewan (then part of the Northwest Territories), making it the oldest federal bird sanctuary in North America. It was, however, a singular event, soon forgotten by all but a few. Then, in 1915, as the U.S.-Canadian treaty was under negotiation, the government decided to set aside twelve lakes in Saskatchewan and fourteen in Alberta as possible future waterfowl refuges, as well as three sites along the St. Lawrence Gulf (Quebec) for shorebirds and one site in Nova Scotia for geese.[87] The three along the St. Lawrence Gulf—Percé Rock, Bonaventure Island, and Bird Rocks—took on a symbolic importance out of proportion to their small size, for they were saved just before the Department of Marine and Fisheries would have destroyed the rookeries there on the (unfounded) grounds that cormorants and gannets were eating all the fish at the expense of the local fisheries.[88] Earlier, Jack Miner had established a small private goose sanctuary on a pond next to his brick factory near Kingsville, Ontario, in 1904, which attracted a lot of press attention and tourists. It quickly became a staging post for thousands of geese on their spring migration to the north, and it was eventually incorporated into Canada's national refuge system.[89]

In 1919, Ottawa updated its enabling law (a full decade before the United States passed its corresponding law), adding a section entitled "Regulations for the Control of Bird Sanctuaries" that laid out a framework for the establishment of federal reserves and for their administration by the commissioner of Dominion parks. As of 1933, the country had established fifty-four bird refuges under Dominion and provincial control, distributed as follows: British Columbia (four), Alberta (ten), Saskatchewan (fourteen), Ontario (one), Quebec (twenty-two), New Brunswick (two), and Nova Scotia (one).[90] The refuges in British Columbia, Alberta, and Saskatchewan were especially important, for nearly three-quarters of Canada's migratory ducks spend their summers in those provinces.[91] As of 1948, the number of reserves had grown to seventy-four, covering 1.15 million acres.[92] By 1996, the number had risen to ninety-eight, covering nearly 28 million acres.[93]

A few Canadian parliamentarians remained hostile to bird protection, and in 1934, they made a last-ditch attempt to abrogate the treaty on the grounds that it favored U.S. sportsmen over their Canadian peers. The Canadian government, however, refused to budge, stating, "If there were no treaty or no regulations these birds which to a great extent breed in Canada might be slaughtered indiscriminately by our friends to the south, and with the exception of a general protest we would have nothing to say in the matter."[94]

Toward a Treaty with Mexico

Negotiating a bird treaty with Canada turned out to be a relatively easy matter. Federal authorities on both sides of the border were supportive of a treaty; the two countries shared a common political, economic, and cultural heritage; and state and provincial game laws were not all that different from each other. Negotiating a treaty with Mexico was far more problematic.

For Mexicans, the name Miguel Angel de Quevedo—"El Apóstol del Arbol" (meaning "The Tree Apostle")—was all but synonymous with conservationism during the first half of the twentieth century. In 1917, Quevedo convinced President Venustiano Carranza (1917–20) to create Mexico's first national park, Desierto de los Leones. He also made sure that the Mexican Constitution of 1917 included a strong conservation clause (Article 27): "The nation shall always have the right to impose on private property the rules dictated by the public interest and to regulate the use of natural elements, susceptible to appropriation so as to distribute equitably the public wealth and to safeguard its conservation." When President Lázaro Cárdenas (1934–40) made environmental protection a linchpin of his administration, Quevedo was called on to create and administer the Department of Forestry, Fish, and Game (later folded into the Department of Agriculture). Under his tutelage, Mexico established forty national parks in five years (thirty-one of which have survived), representing about two-thirds of Mexico's current forty-seven national parks and nearly 85 percent of total park territory (1.6 million of 1.9 million acres, by the late twentieth century). Though best known as a champion of forests, Quevedo worked closely with Edward Goldman of the U.S. Biological Survey to protect birds, successfully lobbying in 1932 for a ban on the use of *armadas* (shooting batteries) in duck hunting, a common practice in the Valley of Mexico. As head of the Mexican Committee for the Protection of Wild Birds (an affiliate of the Audubon Society's International Committee for Bird Preservation), he also championed the broadening of the U.S.-Canadian migratory bird treaty to include Mexico.[95]

Within the United States, the most vocal champion of a Pan-American bird treaty was John H. Wallace, the commissioner of the Alabama Department of Game and Fish. "The United States should take the lead," he told President Wilson upon hearing of the Root and McLean Senate resolutions of 1913, "in proposing an International program under which the migratory wild life of the Western Hemisphere will be accorded the greatest possible protection."[96] Wallace's plea fell on deaf ears at the time, but he tried again in 1919, after the treaty with Canada had been successfully negotiated. "It is clear to my mind," he told Acting Secretary of State Frank L. Polk, "that no program for the conservation of migratory wild life can be effective unless the migratory birds are protected throughout their entire line of migration. To protect them in Canada and the United States, and to permit their slaughter in the Spanish American Republics imposes a tax upon the breeding stock of migratory wild life that it cannot stand without being depleted within the next few years to the point of extinction."[97] This time, he had the support of Senator John H. Bankhead (D-AL), who secured passage of a Senate resolution in February 1920, requesting the president to extend the 1916 Convention to Mexico, Central America, and South America.

The State Department, however, concluded that the timing was not yet right for an agreement with Mexico and that an omnibus treaty with the other Spanish-speaking states, excluding Mexico, made little sense from a political or conservationist perspective. Relations between the two countries, which had been good under the dictatorship of President Porfirio Díaz (1876–1911), had taken a turn for the worse with the outset of the Mexican Revolution in 1910. "I fear that, should we propose such a treaty," one State Department official noted candidly, "Mexico might suggest that we have treaties now existing, which we have chosen to consider as dead letters (for instance, the Extradition Treaty), and that it would be best to restore the functions of such treaties before any attempt should be made to negotiate new ones."[98] Reading between the lines, he was indicating that the U.S. government did not want to have to turn fugitives who had fled to El Paso and other border towns over to the revolutionary Mexican government. It was better to let sleeping treaties lie.

The Department of Agriculture belatedly came to the same conclu-
sion. Spurred initially by the Bankhead resolution, the Biological Survey
sent two of its bird experts—Alexander Wetmore and Edward Goldman—
to South America and Mexico on fact-finding expeditions. Wetmore left
in May 1920 on a yearlong mission to Argentina, Brazil, Uruguay, Paraguay, and Chile to study migratory flight patterns and to build support
for a treaty. "The outcome of this work," as the secretary of agriculture
later explained to the secretary of state, "was that for the present it did not
appear that treaties with those countries would produce practical results
warranting their negotiation."[99] Goldman was sent to Mexico in 1925 to locate the country's principal waterfowl staging posts in order to determine
whether a bilateral bird treaty "would be desirable."[100] He too came back
empty-handed, and he would not return to Mexico to finish his scientific
investigations for nearly a decade.[101]

Progress in the direction of a treaty finally came under the presiden-
cies of Emilio Portes Gil (1928–30) and Lázaro Cárdenas (1934–40). Gil
had become well versed in conservationist issues while serving as Mexico's
secretary of the interior under President Plutarco Elías Calles (1924–28),
and he knew that wildlife issues played well in the United States. In a letter
to the State Department that initiated the negotiations, Manuel C. Tellez,
Mexico's ambassador to the United States, wrote, "It would be of great advantage for our respective countries to take certain measures for the protection of the animal and vegetative marine life and land fauna." Gil and
Tellez had an omnibus "hunting and fishing" treaty in mind, designed to
protect sea life and land fauna, end the smuggling of illicit animal products, impose a duty on hunters and fishers, eradicate predatory animals
(agricultural and pastoral pests), and "protect migratory birds and those
that are useful to agriculture, woodland and public health."[102] Mexico's
newfound desire for a nature-protection treaty dovetailed well with President Franklin Roosevelt's newly initiated Good Neighbor Policy, and the
result was the Convention between the United States of America and the
United States of Mexico for the Protection of Migratory Birds and Game
Mammals of 1936 (hereafter the 1936 Convention).[103]

The 1936 Convention was modeled almost entirely on the 1916 Convention, though there were some notable differences. Negotiators, for instance, no longer felt the need to justify bird protection primarily on the
animals' usefulness to farmers. They also employed the rhetoric of "rational use" rather than species preservation. Article I justified the 1936 Convention on the straightforward grounds that it was "right and proper" to
protect migratory birds so that they would "not be exterminated." It called

for "adequate measures which will permit a rational utilization of migratory birds for the purposes of sport as well as for food, commerce and industry."[104] Article II offered year-round protection for insectivorous birds (eggs and nests included), using language similar to that found in the 1916 Convention. For game birds, however, it established a four-month open season, as compared to the three-and-a-half-month season that bound the United States and Canada. Moreover, unlike the 1916 Convention, it placed no ban whatsoever on spring and summer shooting, except in the case of wild ducks, which could not be shot from March 10 to September 1 (the same dates as in the 1916 Convention). This arrangement left the Mexican government free to allow hunting during the breeding season for all game birds except ducks. It also meant—theoretically at least, though never in practice—that game birds could be under fire for nearly ten months of the year as they flew back and forth between Canada and Mexico.

Article II went slightly beyond the terms of the 1916 Convention in two areas. First, it explicitly called for "the establishment of refuge zones in which the taking of such birds will be prohibited." Second, it prohibited "hunting from aircraft."[105] Neither of these new clauses, however, did much to strengthen the treaty. The refuge clause would have been much more effective if it spelled out in greater detail the number, type, size, and location of these zones (preferably using Canada's system as a model) so as to better guarantee a wide range of avian habitats. For the hunting-methods clause to have had clout, it would have needed to include bans not just on the use of aircraft but also on other "unsportsmanlike" hunting practices—such as site baiting, the use of sink boxes, and the deployment of live decoys—all of which were already banned in the United States and Canada by domestic law. It would also have been useful to ban explicitly the use of armadas in the treaty. Though this practice was already prohibited in Mexico, enforcement was subject to the whims of governmental authorities.

Article III outlawed the transport of migratory birds, dead or alive, across the U.S.-Mexican border. It was similar to the proscriptions established in the 1916 Convention and in conformity with general international trends against the international meat-and-feather trade. It did not, however, restrict in any way the commercial trade in feathers within the territory of Mexico (a practice already banned in the United States). Nor did it regulate in any way commerce in caged birds—a common sight in Mexican markets—even though many of these captured birds (robins, orioles, scarlet tanagers, and waxwings among them) were U.S.-Mexico migrants. Article V did, however, extend the ban on international trade to game mammals. This step was an innovative attempt to restrict the commerce in big-game hunting in anticipation of a more stringent treaty in the future. (It also explains why the official title of the treaty extends beyond migratory birds.)

Article IV listed the migratory game and nongame birds covered by the terms of the treaty. Regarding game birds, the list was practically identical to that of the 1916 Convention (the families Anatidae, Gruidae, Rallidae, and Columbidae and the order Limicolae), except that birds within the order Limicolae were now enumerated under their family names: Charadriidae (plovers), Scolopacidae (sandpipers), Recurvirostridae (avocets and stilts), and Phalaropodidae (highly aquatic shorebirds). As for nongame birds, the treaty protected twenty-three families, including cuckoos, nightjars, swifts, hummingbirds, woodpeckers, flycatchers, wrens, and many others. The list of protected birds mostly overlapped with that of the 1916 Convention (which listed nongame birds by species, not families), but the new treaty extended protection to some new species, including horned larks, blackbirds, grackles, cowbirds, mockingbirds, thrashers, finches, phainopeplas, and sparrows.[106]

Missing from the 1936 Convention was any protection for birds solely on the basis of their rarity or endangerment. Whereas the 1916 Convention gave year-round protection to several disappearing species—including gulls, terns, herons, bitterns, band-tailed pigeons, cranes (little brown, sandhill, and whooping), curlews, and most shorebirds—the newer treaty did not. Also missing from the list were a number of species (mostly birds of prey and fish-eaters) that had become endangered since 1916, including eagles, hawks, owls, kingfishers, pelicans, cormorants, anhingas, ibises, and ravens. A more forward-looking treaty would have included special protections for these species. Article IV, however, did allow for the addition of new birds to the protected list "by common agreement" of the two countries.[107] (This clause would finally bear fruit in March 1972, when the United States and Mexico signed the Agreement Supplementing the Convention of February 7, 1936 for the Protection of Migratory Birds and Game Mammals, which brought an additional thirty-two families under protection.)[108] Also missing from the treaty was an exemption clause that would have allowed Mexico's indigenous populations to engage in subsistence hunting throughout the year. (This defect was not corrected until the Protocol with Mexico Amending Convention for the Protection of Migratory Birds and Game Mammals was signed in 1997.)[109]

The 1936 Convention is best viewed as an extension of the 1916 Convention rather than as a breakthrough treaty in the field of conservation diplomacy. With rare exception (such as the ban on hunting from airplanes), it merely reiterated the U.S.-Canadian treaty. Had it been negotiated twenty years earlier, it would probably have looked only slightly different; agricultural concerns would no doubt have been in the spotlight, and birds would have been listed by species instead of families, but those are the only differences. Conservationists were disappointed with the treaty, mostly because they wanted more birds to be placed under protection, a ban on caged birds, and tighter hunting laws. Judged from the vantage point of 1936, the treaty was indeed disappointing, chiefly because it paid too little attention to habitat protection at the very moment that the United States and Canada had finally committed themselves to the establishment of bird preserves. However, as a State Department spokesperson told treaty critics in 1936, the alternative was not a stronger treaty but no treaty at all: "The Treaty with Mexico only places a minimum as to the protection we shall grant migratory birds in this country. We got the most assurances we could out of Mexico. The laws that they agreed to enact and enforce are certainly more than they have had before in Mexico."[110]

The 1936 Convention was negotiated and ratified in the heyday of the Cárdenas-Quevedo years. Conservationism, however, fared less well under Cárdenas's successor, President Manuel Avila Camacho (1940–1946), who reauthorized the practice of using armadas for duck hunting in the Valley of Mexico. President Miguel Alemán Valdés (1946–1952), an avid sports hunter, took the treaty more seriously: he secured passage of the Federal Hunting Law of 1952, which outlawed the use of poisons to hunt animals, halted the exportation of animal parts, banned shooting in national parks, and established a federal licensing system. His administration, however, did not hire enough game wardens to enforce these measures. Furthermore, none of Cárdenas's successors set up bird reserves along the Gulf and Pacific coastlines, the principal flyways and wintering sites for waterfowl.[111]

American conservation groups rightly deplored Mexico's lax game laws, but the chief beneficiaries of this laxness were U.S. tourists and hunters, not Mexican citizens. Most Mexicans did not eat wildfowl, greatly preferring the taste of chicken and (if they hunted) venison. Commercial duck hunting took place on a significant scale in only two parts of the country—in the Valley of Mexico and along the U.S.-Mexican border—with nearly all of the kill going to restaurants that catered to the tourist industry in Mexico City and the border towns.[112] An investigation by the U.S. Fish and Wildlife Service in 1952, for instance, revealed that Mexico had fewer than five thousand waterfowl hunters as compared to 2 million in the United States and that the total annual Mexican duck kill amounted to around three hundred thousand, less than the number of ducks killed on the *opening day alone* of the U.S. duck season.[113] "I have often thought that were I a duck from the northern United States, Canada or Alaska, when September arrived I would climb to about a thousand feet altitude and head for Mexico," wrote George B. Saunders, the waterfowl expert who assembled the data:

> En route across the United States I would stop at several favorite wildlife refuges, but my long-range compass would be set for Mexico. The destination there would be Tamaulipas, Veracruz, Sinaloa, or any of the other coastal states. For in each there are fine waters where the two-legged duck hunter is a rare or unknown animal, and I could enjoy safety and pleasant feeding until time to migrate northward again. Some will question the wisdom of a duck going to Mexico because of tall tales they have heard or read about armadas (batteries of guns), overshooting, and poisoning of waterfowl there. Of the several million ducks that usually winter in Mexico, most would take that country anytime in preference to the United States, for they are relatively safer there during the open season than they are in the United States during the closed season.[114]

Beyond North America: A Western Hemisphere Treaty?

Efforts to extend treaty protection to migratory birds beyond the territories of Canada, the United States, and Mexico met with mixed success. In December 1935, at the request of the Audubon Society's International Committee for Bird Preservation, the U.S. State Department sent a diplomatic memo to all Central and South American governments soliciting a list of their existing bird-protection laws and enforcement mechanisms. Their responses made clear that most of them lacked the legal machinery upon which a bird treaty could be based. Barbados, Bolivia, Colombia, Costa Rica, El Salvador, Haiti, Honduras, Nicaragua, Panama, Peru, and Venezuela all reported that they had, for all intents and purposes, no laws or restrictions on the hunting and taking of birds. Chile, Cuba, Puerto Rico, Dominican Republic, and Ecuador reported that they had a modicum of laws. Only Argentina, Brazil, and Uruguay claimed to have an extensive number of laws and regulations. Law enforcement, to the extent that it existed at all, was sporadic. Not a single one of the responding states had any bird reserves.[115]

By the end of the 1930s, a new opportunity for bird protection emerged with the signing of the Convention on Nature Protection and Wildlife Preservation in the Western Hemisphere in 1940 (hereafter the 1940 Western Hemisphere Convention). The signers included the United States, Mexico, ten of South America's thirteen states (excepting Guyana, Surinam, and French Guiana), and eight of Central America's eleven states (excepting Jamaica, Belize, and Honduras). The idea for this convention emerged from the Eighth International Conference of American States, which met in December 1938 in Lima, Peru, and it was negotiated under the aegis of the recently created Pan American Union. An equally important force behind the convention was the American Committee for International Wildlife Protection, a committee that included John Phillips, Alexander Wetmore, and other well-known bird conservationists.[116]

The 1940 Western Hemisphere Convention was modeled almost entirely on the 1933 London Convention. It called for the establishment of national parks, national reserves, nature monuments, and strict wilderness reserves "to protect and preserve in their natural habitat representatives of all species and genera of their native flora and fauna, *including migratory birds,* in sufficient numbers and over areas extensive enough to assure them from becoming extinct through any agency within man's control." Article I defined the phrase *migratory bird* broadly to include "birds of those species, all or some of whose individual members, may at any season cross any of the boundaries between the American countries," including Charadriidae (plovers), Scolopacidae (sandpipers), Caprimulgidae (nightjars and whippoor-wills), and Hirundinidae (swallows). Article VII, the only one exclusively dedicated to bird protection, stated in full: "The Contracting Governments shall adopt appropriate measures for the protection of migratory birds of economic or aesthetic value or to prevent the threatened extinction of any given species. Adequate measures shall be adopted which will permit, in so far as the respective governments may see fit, a rational utilization of migratory birds for the purpose of sports as well as for food, commerce, and industry, and for scientific study and investigation."[117]

The treaty encouraged each country to submit a list (to be attached to the treaty as an annex) of all flora and fauna that it considered endangered enough to deserve special protection within its territory. Unfortunately, no country was even obligated to prepare such a list, let alone to protect any of the species it listed. Nor was there any incentive to work in conjunction with neighboring states to jointly protect endangered species. As a consequence, a patchwork of bird-protection regulations emerged, reminiscent of what had prevailed among the U.S. states before passage of the

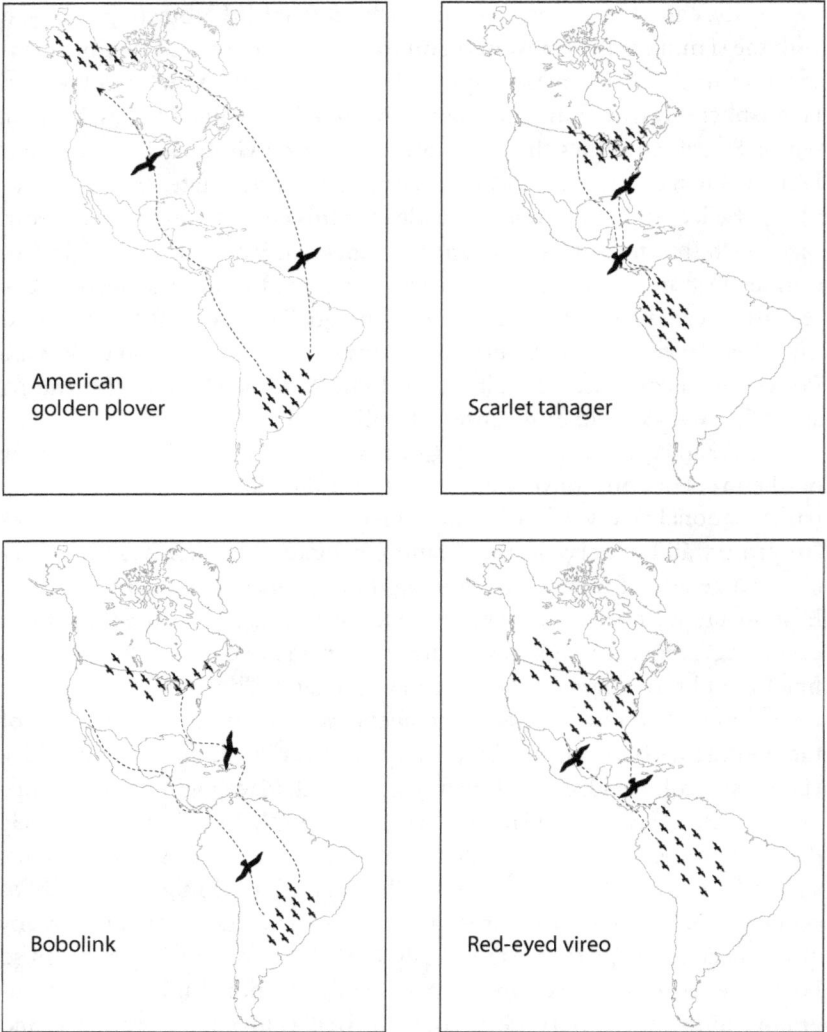

Figure 2.5. Migratory patterns of the American golden plover, scarlet tanager, bobolink, and red-eyed vireo. None of these species remains within the confines of the 1916 and 1936 treaties during their annual migrations. Adapted from Jean Dorst, *The Migrations of Birds* (Boston: Houghton Mifflin, 1962), 110, 114, 123, and 137.

Weeks-McLean Law in 1913. Brazil, for instance, placed a closed season on the hunting of nine families (Tinamidae, Anatidae, Cracidae, Phasianidae, Rallidae, Cariamidae, Scolopacidae, Recurvirostridae, Columbidae) and extended year-round protection to another fifty-two. Neighboring Bolivia, by contrast, submitted a list that covered just thirteen species (the house wren, ovenbird, swallow, ostrich, tero-tero, plover, seagull, calandra lark, magellanic woodpecker, owl, great kiskadee, the little blue heron, and white heron), and Paraguay submitted no list at all. Similarly, Argentina extended protection to twenty-two bird species, but its neighbor Chile did not bother to compile a list. In the Caribbean, Cuba produced a long list of protected species, whereas the Dominican Republic and Haiti produced only short lists. In Central America, El Salvador offered protection to two large orders, Passeriformes ("perching birds") and Falconiformes (hawks and vultures), as well as to seven families and five species. But Nicaragua listed only three families, Guatemala just twenty-eight species, and Ecuador a mere nine species.[118]

Efforts to use the 1940 Western Hemisphere Convention to jump-start national parks and wildlife reserves brought a measure of success. A few countries, such as Argentina, Venezuela, and Mexico, had already established some parks and reserves before 1940, and the new treaty spurred them to develop more. Other Central and South American countries slowly followed suit—sometimes after decades of delay—and by the end of the twentieth century, there were hundreds of parks, reserves, wilderness areas, and other protected regions for birds and other animals. A few, notably Costa Rica, have made the parks and reserves a major source of tourist dollars. More typically, however, states have been lax in enforcing game laws and in shielding the protected areas from outside development and encroachments.

"In the United States there are now out hunting, in this very season, 48 *big armies* of men," Hornaday wrote in his aptly titled retrospective, *Thirty Years War for Wild Life* (1931). "Their grand total strength," he said, "is about 7,500,000 well armed, well equipped, and money-supplied killers of 'game' and pseudo-game. This means 7,500 *regiments of full strength!* The grand total is composed of 6,493,454 licensed hunters, plus about 1,500,000 unlicensed hunters who legally hunt local game on their own lands without licenses. It far exceeds in number *all of the active standing armies of the world.*" He added: "The progressive extinction of all United States game and near-game birds is rapidly proceeding. Ninety percent of it is due to merciless and determined shooting; and we greatly fear that the leaden-footed Big Stick of the Law will not overtake the 48 huge armies of killers before the game takes its final plunge into oblivion."[119]

Nearly twenty years had passed since the publication of *Our Vanishing Wild Life,* but Hornaday still viewed hunters in the forty-eight states of the continental United States as a vast Army of Destruction. Like the proverbial generals of World War I, he was fighting the last war rather than preparing for new ones on the horizon. Lead-footed or not, the U.S. and Canadian governments had successfully ended the bird carnage that had marred the previous century. The 1916 Convention (and the later 1936 Convention) went a long way toward weaning North Americans from their excessive reliance on nature's bounty. The era of commercial hunting—once the single greatest threat to the flyways and rookeries of North America—came to an abrupt end. To survive, milliners turned to ribbons and bows instead of bird parts, and meatpackers turned increasingly to farm-produced poultry (chicken and turkey farms) instead of wildfowl.[120] Recreational hunting was still going strong, but it was now a highly regulated sport, held in check by closed seasons, bag limits, and license requirements. As international hunting accords, the 1916 and 1936 treaties have to be judged successes: no North Americans today dine regularly on robin soup and goldfinch pie, adorn themselves with heron aigrettes and hummingbird heads, or consider it sporting to wipe out an entire flock and rookery in a single day.

As habitat-protection accords, the treaties were far more problematic, largely because they did not provide an adequate framework for protecting breeding, feeding, and wintering sites along the principal North American flyways. Hornaday, for instance, insisted that there were thirteen bird species on the brink of extinction in 1931, five of them game birds—the woodcock, sharp-tailed grouse, pinnated grouse (prairie chicken), sage grouse, and golden plover—and he predicted the Army of Destruction would soon blast all five to oblivion.[121] Some eighty years later, however, not a single one has become extinct, and among those that are today still endangered (notably the pinnated grouse), habitat loss rather than overhunting is the principal cause. The same is true of ducks, geese, and many other game birds that were not included in Hornaday's list: the gradual decline in their numbers since 1916 has been far more closely linked to the disappearance of marshes and other wetland habitats than to the impact of hunters.

One major problem is the geographic limitations of the 1916 and 1936 treaties. The combined geographic areas of Canada (3.85 million square miles), the United States (3.54 million), and Mexico (756,000)—though totaling over 8.1 million square miles—are not large enough to contain the entire habitat range of many migratory bird species. Birds on the Pacific flyway are relatively well protected because they characteristically breed in Canada and Alaska; utilize the western half of Canada and the United States as a sky route; and winter in the United States, Mexico, and (in some cases) Central America. The Canada goose, Ross's goose, gray-cheeked thrush, fox sparrow, and Connecticut warbler are all examples of birds whose ranges are largely confined to Canada and the United States. The mourning dove and common snipe are examples of birds whose range is largely coterminous with Canada, the United States, and Mexico (plus Central America). But other birds—especially those that utilize the Mississippi and Atlantic flyways—are far more exposed to predation: many of them fly to South America via the Caribbean, skipping Mexico altogether or utilizing only the Yucatan. The black-poll warbler, cliff swallow, black-and-white warbler, redstart, rose-breasted grosbeak, and red-eyed vireo all range between Canada and the northern half of South America, whereas the golden plover, bobolink, and scarlet tanager range between Canada and the southern half of South America. The Arctic tern travels from Canada to Europe to Antarctica and then back again to Canada.[122]

Recognizing some of the geographic limitations, the United States signed separate bird treaties with two additional countries in the eastern Pacific: the 1972 Convention for the Protection of Migratory Birds and Birds in Danger of Extinction, and Their Environment, signed between the United States and Japan, and the 1976 Convention between the United States of America and the Union of Soviet Socialist Republics Concerning the Conservation of Migratory Birds and Their Environment. Japan subsequently signed two other bilateral bird treaties, one with Australia and the other with the Soviet Union, further extending the network. However, these additional agreements, though offering extra protection to birds that utilize the eastern Pacific Rim as part of their migratory routes, were never extensive enough to offer full protection to most bird species. Many more countries would have had to sign bilateral or multilateral arrangements for this approach to be effective.

Wetland drainage also proved a major hurdle to effective bird protection. The greatest loss, by far, has occurred along the flyways of the continental United States, where agricultural and urban development has resulted in a large amount of habitat loss. In 1780, the area that now constitutes the United States contained 392 million acres of wetlands, of which around 221 million acres were found in the forty-eight continental states, 170 million acres in Alaska, and 59,000 acres in Hawaii. As of 1980, the continental United States had lost 117 million acres (a 53 percent drop) and Hawaii around 7,000 acres (a 12 percent drop), whereas Alaska had retained almost all of its wetland acreage (less than a 1 percent drop). Florida had taken the hardest hit, losing 9.3 million of its original 20.3 million acres (a 46 percent loss), followed by Texas, which lost 8.39 million of its original 16 million acres (a 52 percent loss), and Louisiana, which lost 7.41 million of 16.2 million acres (a 46 percent loss). Meanwhile, California had lost 91 percent of its original wet-lands (4.55 million out of 5 million acres), followed by Ohio with a loss of 90 percent (4.52 million out of 5 million acres), and Iowa with a loss of 89 percent (3.58 million out of 4 million acres).[123] This massive depletion of wetland acreage has far outstripped the acres put aside as bird reserves and national parks over the past hundred years, and the end result is that birds now have less, not more, space on which to land.

Habitat loss in Canada was far less severe (with wetlands still cover-ing 14 percent of its territory), though agricultural drainage in certain key regions has had a dramatic impact on some bird species. Waterfowl popu-lations were particularly hard hit by the gradual drainage of "prairie pot-holes" (shallow pools ranging from an acre to hundreds of acres in size) in Alberta, Manitoba, and Saskatchewan to make room for farmland. By the 1970s, Alberta had lost 61 percent of its original wetland space, and southwestern Manitoba experienced a 57 percent decline. Damage in Sas-katchewan was less severe, but nonetheless, 84 percent of its wetlands had been affected in one way or another by human activities. On the west coast, British Columbia had lost 70 percent of its wetlands to agriculture, much to the detriment of shorebirds. On the east coast, the Bay of Fundy on the Nova Scotia–New Brunswick border has been hard hit by development. Ontario and Quebec have also experienced huge wetland losses, especially in the Great Lakes region and the St. Lawrence Valley, prime sites for mi-gratory birds.[124] As in the United States, there has been a net loss of wetland acreage, despite the proliferation of bird reserves.

The lack of reliable data on Mexico makes it much more difficult to ascertain the rate of habitat loss there, but the combined impact of agricul-tural development, cattle ranching, industrial pollution (especially from oil, chemicals, and petrochemicals), deforestation, and tourism must have been severe. Equally problematic is the fact that only five of the country's sixty-six major wetlands regions are under protection today. Wetland drainage as a consequence of agricultural irrigation and river impoundments along the Pacific and Gulf coasts—prime wintering spots for waterfowl and other migrants—has been particularly intense since the mid-1980s. Oil extraction, meanwhile, increasingly threatens the wetlands of Tabasco and Vera Cruz. The waterfowl-rich wetlands in the interior highlands (including Chihua-hua, Coahuila, Guanajuato, Jalisco, and Michoacan) have also been affected recently, mostly due to logging, grazing, cultivation, and urban growth.[125]

Still, Hornaday was right about one thing: the 1916 Convention was more a cease-fire than a full-fledged peace agreement. The negotia-tors did not foresee—and indeed could not have foreseen—all of the challenges that lay ahead for bird conservation, from water-drainage projects to oil spills to avian flu. They did comprehend, however, that *commercial* hunting for meat and feathers had become annihilationist, and they managed to stop the bloodletting before irreversible damage had been done to most game species. They accomplished a good deal more in the process, extending protection to hundreds of "agricultural birds" and other species that would normally not be included in a typical hunting treaty. State of the art when it was first conceived, the 1916 Convention quickly became the foundation upon which all bird protection in North America rested, even if it gradually began to show its age and require some updating to keep it functional. Although few negotiators hold up the 1916 Convention as a model for global bird agreements today, it was the only effective bird treaty in the entire world for many decades, and it certainly helped save many North American species from the butcher's block.

Time Line of Migratory Bird Protection

1896   In *Geer v. Connecticut*, the U.S. Supreme Court determined that wild animals were under the jurisdiction of the states, not private landowners.

1900   The Lacey Game and Wild Birds Preservation and Disposition Act (Lacey Act) became law in the United States. It made it illegal to transport birds across state lines if the birds were killed in violation of state law.

1913   The Migratory Bird Act (Weeks-McLean Law) placed migratory birds in the United States under federal jurisdiction. Opponents challenged its constitutionality in the courts.

1916   The Convention for the Protection of Migratory Birds (or the 1916 Convention) was signed between the United States and Canada. It established a framework for creating uniform game laws in both countries and placed some endangered birds under year-round protection.

1917   The Migratory Birds Convention Act was passed by the Canadian Dominion government in August. It established game regulations for Canadian provinces in conformity with the 1916 Convention.

1918   The Migratory Bird Treaty Act was passed in the United States. It brought U.S. federal and state laws into conformity with the 1916 Convention.

1920   In *Missouri v. Holland*, the U.S. Supreme Court upheld the constitutionality of the 1916 Convention. Oliver Wendell Holmes wrote the majority opinion.

1920   In *King v. Russell C. Clark*, the Supreme Court of Prince Edward Island upheld the validity of the 1916 Convention. The presiding judge cited Justice Oliver Wendell Holmes in his decision.

1924   The Upper Mississippi Wildlife and Fish Refuge Act established the first large federal wildlife reserve for birds and fish in the United States.

1928   The Bear River Migratory Bird Refuge Act established a bird reserve near Utah's Great Salt Lake in the United States.

1929   The Migratory Bird Conservation Act created a commission to purchase land for migratory bird reserves in the United States.

1936   The Convention between the United States of America and the United States of Mexico for the Protection of Migratory Birds and Game Mammals (or the 1936 Convention) was enacted. This agreement extended the 1916 Convention to Mexico with minor textual changes.

1937   The Pittman-Robertson Act (Federal Aid in Wildlife Restoration Act) augmented the U.S. federal government's role in wildlife restoration projects, including bird protection.

1940   The Convention on Nature Protection and Wildlife Preservation in the Western Hemisphere was signed. This agreement promoted the creation of nature parks, including bird reserves, throughout the Americas.

1950   International Convention for the Protection of Birds, signed in Paris, became first bird treaty of importance outside of North America.

1966   The Endangered Species Act became law in the United States (updated in 1969 and 1973).

1971   The Convention on Wetlands of International Importance Especially as Water-fowl Habitat was signed in Ramsar, Iran. It was the first multilateral treaty of importance to protect bird habitat.

1972   The Agreement Supplementing the Agreement of February 7, 1936 between the United States of America and the United States of Mexico for the Protection of Migratory Birds and Game Mammals was signed. This pact expanded the list of birds protected under the terms of the U.S.-Mexican treaty.

1972   The Convention for the Protection of Migratory Birds and Birds in Danger of Extinction, and Their Environment, signed between the United States and Japan, partially extended the 1916 Convention to a small portion of the Pacific Rim.

1973   The Convention on Trade in Endangered Species of Wild Fauna and Flora (CITES Convention), signed in Washington, DC, became the first comprehensive attempt to save endangered species worldwide.

1976   The Convention between the United States of America and the Union of Soviet Socialist Republics Concerning the Conservation of Migratory Birds and Their Environment was signed. It extended portions of the 1916 Convention to the Soviet Union.

1979   The Convention on the Conservation of Migratory Species of Wild Animals, signed in Bonn (Bonn Convention), became first treaty of importance to protect migratory birds worldwide.

1986   The North American Waterfowl Management Plan was inaugurated by the United States and Canada to further protect North American game birds.

1995   The Protocol Amending the 1916 Convention for the Protection of Migratory Birds updated several articles of the 1916 Convention and exempted indigenous peoples from the closed-season clauses.

1997   The Agreement Supplementing the Convention of February 7, 1936 for the Protection of Migratory Birds and Game Mammals exempted indigenous peoples from the closed-season clauses of the 1936 Convention.

Chapter 3

The Antarctic Whale Massacre

They say the sea is cold, but the sea contains
the hottest blood of all, and the wildest, the most urgent.

—D. H. Lawrence, "Whales Weep Not!"

NORWAY—THE "Land of the Midnight Sun"—is a magnet for people in search of "unspoiled" nature. It possesses a long and jagged coastline with spectacular fjords and a narrow interior with high and rugged peaks. Hundreds of bird species soar through its skies, and salmon, cod, and capelin teem along its shores. Sparsely populated for its size (4.5 million people in an area of 125,000 square miles), Norway has just a single major city, its capital, Oslo (population 503,000). Cocooned in splendor, Norwegians do not need to drive to a national park to enjoy the great outdoors. Yet this nation also harbors a deeply troubling environmental legacy: from the 1860s to the 1960s, Norway was the greatest whaling power in the world. For the better part of a century, Norwegians killed more whales than did all other peoples on the face of the earth.[1]

The person most identified with Norway's rise to prominence as a whaling power was Svend Foyn. Foyn was born in 1809 in Tønsberg, a port town in the province of Vestfold, on the western side of the Oslo fjord, just south of Oslo. He made his fortune as a sealer on Jan Mayen Island (north of Iceland) before turning his attention to whaling in the 1860s, an endeavor that preoccupied him until his death some three decades later in 1894. A risk-taker and an innovator, Foyn poured his time and money into the improvement of hunting techniques. In 1863, he built the world's first steam-powered whale catcher, *Spes et Fides* (Hope and Faith). Less than a hundred feet long and equipped with a fifty-horsepower engine capable of only seven knots, it merely hinted at the mammoth, oil-powered whale catchers that would come later. But it already far outpaced the traditional sail-driven schooners and rowboats of his day, and it required a much smaller crew, so its design was soon adopted by others. Foyn was also among the first whalers to install an accumulator (a winch-and-rope system that functions much like a fishing rod) on his ship, thereby greatly reducing the frequency of snapped ropes and lost whales. In the 1880s, he installed yet another device, an air-compression pump, which forced air into a captured whale's body cavity to keep it afloat for easy processing.[2]

Foyn's best-known invention—and the one that earned him the nickname "father of modern whaling"—was the harpoon gun. Patented in 1870, it provided whalers with a surefire method for killing all species regardless of their size or speed. Accurate and reliable, it became the model for all subsequent whaling guns for the next hundred years. Shaped like a piece of light artillery, it was mounted on a swivel on the vessel's bow. It consisted of three distinct devices that worked together: a cannon, a harpoon, and a grenade (originally with gunpowder and sulfuric acid). The cannon was used to launch a harpoon deep into the whale's backside. The harpoon was fastened to the ship with a rope. The harpoon tip contained the grenade, and just behind the grenade were four backward-facing barbs. Upon impact, the four barbs opened outward (much like umbrella ribs), preventing the whale from dislodging the harpoon. The movement of the barbs also triggered the exploding device, which detonated once the harpoon was lodged inside the whale's body. If all worked as planned, the whale died within a half hour or so, without pulling and thrashing the ship too much.[3]

Foyn's many innovations gave the Norwegians a competitive edge that they would not fully relinquish until the 1960s, by which time most of the world's whale stocks were already severely depleted. His home province of Vestfold, Norway's most populated region and busiest shipping center, emerged as the center of the world whaling industry. Its three main ports—Sandefjord, Tønsberg, and Lavik—generated much of the capital and produced most of the ships and crews for whaling expeditions worldwide, even those flying under foreign flags. (In 1931, for instance, all but 142 of the 10,691 whalers in the Antarctic were from Norway, of which nearly 8,000 came from Vestfold alone.)[4] Norwegians pioneered in the development of the modern floating factory, the high-pressure boiler, and the stern slipway. They discovered methods for turning whale wastage into usable products such as animal feed and fertilizer. They designed stronger ropes and sturdier harpoons. They were the first to fully exploit the Arctic hunting grounds, the first to open up the Antarctic, and the first to engage in modern pelagic whaling. Norwegian was even the official language aboard nearly all whaling ships, and the one word every whaler understood without a translator was *hvalblast* (meaning "thar she blows").[5]

So thoroughly did Norway dominate the whaling industry after the 1860s that all previous hunting techniques came to be known as "old whaling" and Norwegian ones as "modern whaling." Sail and human muscle powered the old vessels, whereas coal and oil fueled the modern ones. Old whalers used handheld harpoons, but modern ones were grenade-tipped. Old whalers fastened the carcass to the ship's side or hauled it to a nearby shore station in order to strip (flense) the fat (blubber) from the whale. Modern whalers, by contrast, learned how to hoist a dead whale directly onto a ship's deck for quick flensing. Old whalers boiled (rendered) the blubber slowly in specially designed cooking pots (tryworks) and then stored the oil in barrels; at best, they could squeeze only about half the oil from the carcass. Modern whalers learned how to render the entire carcass in high-pressure boilers and store the oil in huge tanks, wasting little. Old-style whaling was an artisanal enterprise, limited in size and scope. Modern whaling was an industrial enterprise, limited only by the number of whales in the ocean. Old whalers stayed mostly in the Arctic regions, close to their shore stations and home markets. Modern whalers worked almost entirely in the Southern Ocean (south of 60° south latitude) around the Antarctic continent. Old whaling led to the gradual depletion of one hunting ground after the next over many centuries. Modern whaling was undertaken on a scale so massive that it can only be called exterminationist.

The figures tell the tale. The Basques, the world's first great commercial whalers, snagged about five hundred whales per year between 1530 and 1610.[6] The Dutch, the foremost whalers of the seventeenth century, killed an average of eight hundred bowheads (their favorite whale species) between 1670 and 1719.[7] By contrast, Antarctic whalers killed around nine thousand whales per year between 1904 and 1924 and over thirty thousand per year in the 1930s.[8]

Whale Biology

Porpoises, dolphins, and whales are all mammals belonging to the order Cetacea. There is no clear biological line separating whales from porpoises and dolphins, except a few minor anatomical characteristics. The porpoise has a rounded snout and a length of four to eight feet, whereas the dolphin has a beaklike snout and a length of eight to twelve feet. Whales are generally over twelve feet in length, and "great whales" are between thirty and one hundred feet long. Physical size, however, is a poor gauge because there are many pygmy whale species that are the same size as porpoises and dolphins. To complicate matters, some common nomenclatures are inaccurate: the melon-headed whale and killer whale, for instance, both belong to the dolphin family. The terms *Cetacea* and *whale* were often used interchangeably in whaling treaties, even when it was clear that the diplomats were solely concerned with the great whales—the only ones regularly hunted. This carelessness gave rise to a still-unresolved controversy as to whether the treaties apply only to the great whales or to porpoises, dolphins, and small whales as well.[9]

Like other mammals, cetaceans are warm-blooded lung breathers that give birth to live young and nurse through milk-filled glands. They are, however, the only mammals that have become completely adapted to water. To keep their vital organs insulated from the frigid oceans, their bodies are surrounded by a thick layer of blubber. Aside from keeping them warm, blubber gives them more buoyancy in water and serves as a food reserve during the breeding season. Their "nose" is a blowhole atop their heads linked directly to their lungs; to breathe, whales break the surface of the water, blow out the air in their lungs, and ingest new air, all within a few seconds. The frigid oceans once gave whales a level of security that no land mammal enjoyed. Their thick layer of blubber, however, made them a coveted prize for human hunters, and their breathing mechanism made them visible to whalers and thus vulnerable to the harpoon.[10]

Cetaceans are divided into two suborders—Mysticeti ("moustached whales") and Odontoceti ("toothed whales")—based largely on their feeding methods. Mysticeti are more commonly known as baleen whales because they trap food in curtains of flat, keratinous rods (baleens) attached to the roof of the mouth. They feed mostly on tiny creatures—zooplankton and small fish—using the baleens to filter out water and debris. Some species skim the surface of the water to capture their prey, and others swim below the surface and swallow in big gulps. The largest of the baleen whales are known as rorquals (derived from the Norwegian *rörhval*, meaning "furrowed whale"). Rorquals have distinctive grooves about three inches deep that begin at their throats and run longitudinally about one-third of the way down their undersides.

The other suborder of whales—Odontoceti—have teeth instead of baleens in their mouths. They prey mostly on fish, squid, and octopus. Most have cone-shaped teeth that help them engulf (not chew) food, but there is considerable variation among different species. Some have two hundred teeth; the male narwhal, however, has just a single long tooth that juts spearlike eight feet or more from its mouth. Feeding methods mark the main difference between Mysticeti and Odontoceti, but there are other distinctions as well. Baleen whales are generally larger than toothed whales, and baleen females are generally larger than their male counterparts. Baleens tend to travel alone or in small groups, whereas toothed whales often prefer to live and move in larger groups. Baleens have paired blowholes, toothed whales only one. Baleens cannot echolocate ("see" with sound), but toothed whales can. Baleens are monogamous, toothed whales generally polygamous. Finally, all members of the Mysticeti suborder are whales, whereas the Odontoceti suborder includes dolphins and porpoises.[11]

Mysticeti and Odontoceti are further divided into nine families and seventy-five species. There are only three Mysticeti families—the Balaenidae (right whales), Balaenopteridae (rorquals), and Eschrichtiidae (gray whales)—but these families include all the species that have been hunted to near extinction over the past two centuries. There are six families of Odontoceti: Physeteridae (sperm whales), Monodontidae (narwhals and belugas), Ziphiidae (beaked whales), Delphinidae (oceanic dolphins), Phocoenidae (true porpoises), and Platanistidae (river dolphins). Among these, only sperm whales have been hunted to any significant extent. Porpoises and dolphins were not spared out of nostalgia or respect but because they were considered too small to be worth the chase. The term *hunted whale* is therefore all but synonymous with *great whale;* except for the sperm whale, it is also synonymous with *baleen whale.*

Whales have torpedo-like shapes and powerful tails (flukes), which they use to propel themselves in water. Nearly all of the great whales undertake an annual migration between polar regions and temperate regions. Most are "cosmopolitan" (that is, they inhabit all of the world's oceans), but typically, they do not cross the equator, so northern and southern populations are isolated from each other. Migrating whales head to the polar seas to feed when it is warmest there—April to September in the Arctic, October to March in the Antarctic. For humans, the polar seas always seem bleak and barren, but baleen whales find a smorgasbord of zooplankton

and fish there during summer. In the Arctic, they rely primarily on ptero-
pod mollusks and various crustacea known colloquially as brit. Antarctica
is even richer in food, especially along the so-called Antarctic convergence,
where cold and warm currents come together. Whales there feast primarily
on *Euphausia superba*, a shrimplike crustacean about two inches in length,
known by the Norwegian word *krill*. Toothed whales prefer fish and squid,
so they are far less dependent on polar feeding, but bachelor sperms are
known to migrate there nonetheless.[12]

During the breeding season—October to March in the Arctic, April
to September in the Antarctic—whales leave the polar regions and head
to tropical and subtropical seas. Gestation and nursing periods vary from
species to species, but rorquals are generally on a two-year reproduction
cycle and right whales on a three-year cycle. Females give birth around
eleven months after mating and typically nurse for another six months.
Sperm whales follow a different rhythm: they gestate for fifteen months,
nurse for two years, and reproduce every four years. These slow reproduction rates made it difficult to protect whale stocks in the face of round-the-
clock hunting, and they have greatly slowed the replenishment rate in the
posthunting era. The fact that female baleens are larger than male baleens
and that pregnant females are the largest (and most oil-rich) of all has also
worked against conservation.[13]

Until the twentieth century, nearly everything that was known about
whales came from the firsthand accounts of whalers and the onboard investigations of amateur scientists, often filtered to the public through the
fertile imaginations of storytellers. William Scoresby's *Account of the Arctic Regions with a History and a Description of the Northern Whale-Fishery*
(1820) and Herman Melville's *Moby Dick* (1851) are among the most famous
and accurate of these accounts. By 1900, many biologists were becoming
increasingly concerned about the long-term prospect for whale conservation in light of modern hunting techniques. At their urging, the British
Colonial Office established the Discovery Committee (headquartered in
Grytviken, South Georgia), which launched a succession of ships—*Discovery*
(1920), *William Scoresby* (1925), and *Discovery II* (1929)—to observe distribution patterns, behavioral characteristics, and reproduction rates. In 1929,
the Norwegian government established three interlocking institutions: the
Hvalråd (Whaling Council), composed of government leaders and scientists; the Statens Institutt for Hvalforskning (State Institute for Whale Research), a center for cetacean science; and the Komitéen for Internasjonal
Hvalfangststatistiskk (International Committee for Whaling Statistics),
which tracked whaling ships around the world. Out of Norway's efforts

emerged the annual *International Whaling Statistics,* still the most reliable and comprehensive source of kill data worldwide. Diplomats relied almost entirely on Britain's scientific reports and Norway's official statistics when they formulated the international treaties of the 1930s and 1940s. Unfortunately, there were still two glaring gaps. Too little was known about reproduction patterns, and therefore, the treaties allowed for the taking of many juveniles that were mistakenly thought to be fully grown. Too little was also understood about migration routes, and as a consequence, the same populations were allowed to be targeted at different sites during various times of the year.[14]

Until Foyn invented the grenade harpoon, only five species of great whales—rights, bowheads, humpbacks, grays, and sperms—were hunted to any significant extent. These were, for the most part, slow-moving (around seven knots per hour) and coast-hugging species that could be caught readily with premodern hunting methods. Rights and bowheads were greatly preferred because they floated after dying, making them easier to secure to a boat and haul to shore. Humpbacks and grays were less desirable because they did not yield as much oil per pound as rights and bowheads. Moreover, humpbacks (like all rorquals) sank when they died, and grays had the reputation of being fierce fighters. As for sperm whales, only U.S. whalers built an entire industry around killing them; others chased them only occasionally.

With the advent of Norwegian-style whaling in the 1860s, five more whale species (all rorquals) became targets: blues, fins, seis, Bryde's, and minkes. They were either too large, too fast (moving at up to thirty knots per hour), or too elusive for earlier generations of whalers to catch with any regularity. Unfortunately, modern whalers did not stop hunting the first five species when they began chasing the next five; on the contrary, the new hunting techniques just made it all that much easier for them to mop up the remnant populations of rights, bowheads, grays, and humpbacks; these species thus became commercially extinct in many parts of the world long after they were the prime targets, casually killed by gunners in search of the more highly prized rorquals.

Right Whales

The right whale—also sometimes known as the nordcaper and North Cape—got its common name from British whalers back in the days when it was the most coveted species and hence the "right" one to pursue. There are two distinct populations, one that remains in the Northern Hemisphere (*Balaena glacialis*) and one that remains in the Southern Hemisphere

(*Balaena australis*). Rights reach a maximum length of only sixty feet and a weight of eighty tons, but they produce more oil per pound than do most other whales. They are black or gray, except for a white patch on their undersides, and they have keratinous lumps, known as callosities, on their heads. Their lower jaws are large, and their mouths contain two large baleen plates, each eight to ten feet long. Like most whales, rights feed on zooplankton in the cool polar regions, then migrate each year to warmer and shallower waters to breed and give birth.

Bowhead Whales

The bowhead whale (*Balaena mysticetus*)—also known as the Greenland whale and polar whale—is a larger and rounder cousin of the right, found only in the Arctic regions. It stays year-round in cooler waters, never migrating southward to temperate waters. Black except for a white patch on its lower jaw, it reaches a maximum length of sixty-five feet and a weight of sixty-five tons. Its huge head contains the longest baleen plates of any whale, up to fifteen feet in length. For that reason, it was the whale of choice whenever women's fashions called for corsets or hoops. Bowheads were once plentiful throughout the Arctic region, but Dutch and British whalers all but eliminated the eastern Arctic populations in the eighteenth century, and U.S. whalers decimated the western Arctic stocks a century later. Today's remnant population spends its winter months in the Bering Sea, then migrates northward toward the Arctic pole in April and May.

Humpback Whales

The humpback whale (*Megaptera novaeangliae*) is the only slow-moving rorqual and thus the only one that Europeans hunted before the advent of modern whaling. Its baleens are short (about two to three feet long) and its oil yield meager as compared to rights and bowheads, so it was rarely the preferred target. A thickset creature, its name probably comes from the fact that it arches sharply when it dives. Typically, it is black on top and white on the bottom, and it attains a length of fifty to sixty feet and an average weight of forty tons. Humpbacks are the delight of whale watchers today because three-quarters of their bodies come out of the water when they breach and because the males make elaborate sound patterns that resemble human song. Groups of humpbacks sometimes engage in a unique ritual known as "bubble-net feeding": they surround a school of fish and then release air bubbles, which force the fish more tightly together; the whales then lunge in and swallow their prey.[15]

Gray Whales

The gray whale (*Eschrichtius robustus*) is nicknamed the "devil fish" because females with young have been known to ram ships when harassed or harpooned. Never as numerous as other whale species, grays once inhabited the Atlantic and Pacific oceans, but humans eradicated them from the Atlantic centuries ago and decimated their numbers in the western Pacific more recently; consequently, their distribution is now almost entirely limited to the eastern Pacific from Alaska to Mexico (and this too is a remnant population that barely survived the slaughter of California whalers). Along with humpbacks, the gray is a favorite of whale watchers. It reaches a length of forty to fifty feet and thirty-five tons in weight. Aside from its characteristic color, the gray can be identified by the absence of a dorsal fin and by the presence of "knuckles" (crenulations) on its backside; it is also the only large whale with an upper jaw that overhangs the lower one. The gray makes an annual migration from its feeding grounds in the Bering and Chukchi seas to Baja California, a distance of more than seven thousand miles, the longest known migration of any mammal in the world. The gray is also unusual in that it is a bottom-feeder: it dives to the ocean floor and skims the bottom for copepods, using its baleen plates to filter out mud and water.

Sperm Whales

The sperm whale (*Physeter macrocephalus*)—also known as the cachalot and pottwal—is the only great whale belonging to the Odontoceti suborder. It is black in color, except for a white lower jaw and a white patch on its underside. Among the great whales, it is the only polygamous species: dominant males impregnate many females, and the rest ("bachelors") are pushed to the sidelines. Males weigh up to sixty tons and reach sixty feet in length; females are about one-third smaller. The sperm whale's enormous square head contains the world's largest brain, the world's largest breathing apparatus, a row of conical teeth, and a compartment ("case") filled with a clear amber liquid known as spermaceti. Like other Odontoceti, the sperm whale locates its prey (mostly small squid) by echolocation; it may also stun them with sonic blasts before swallowing them. U.S. whalers targeted sperm whales for well over a century (roughly from 1720 to 1860) without seriously depleting their numbers, probably because they preferred the larger males over the smaller females. Sperms were not regularly hunted again until after 1945 (and especially after 1962), as other species became depleted.[16]

Blue Whales

The blue whale (*Balaenoptera musculus*) is the largest animal ever known to have lived: it can reach one hundred feet in length and weigh one hundred and fifty tons or more. It has a greenish-blue color, but its underside is often laden with diatoms, giving it a yellowish hue and the nickname "sulphur bottom." Blues are found worldwide, but they exist as several distinct populations that do not commingle, even when they feed, breed, and migrate in neighboring waters. They are not particularly gregarious, preferring to travel alone or in small groups. Their distinctive color and enormous size, as well as rounded upper jaw, make them easy to distinguish from other whales. A subspecies of the blue, the pygmy blue (*Balaenoptera musculus brevicauda*), is found only in the Southern Hemisphere. Its name is doubly deceptive: it attains a length of eighty feet (small only by blue whale standards), and it is silvery-gray in color. Like other rorquals, the blue is an aggressive feeder: it opens its jaws wide and uses its lower jaw to scoop krill (its preferred food) into its mouth. During its four-month feeding season, a blue whale consumes around 1.5 million calories per day. Whalers hunted the Antarctic blues to the brink of extinction in the first half of the twentieth century and most intensely between 1913 to 1937. During the 1930–31 season alone, they took the largest number of blues ever recorded—29,410 according to the official tally—and over the following thirty-five years, they took another 189,710 blues, for an average of 5,420 per year.[17] To this day, the stocks have never recovered from this onslaught.

Fin Whales

The fin whale (*Balaeoptera physalus*), also known as a finner and razorback, is seventy-five to eighty feet in length and fifty tons in weight, making it second only to the blue among the earth's largest animals. As its name implies, it has the most pronounced dorsal fin of any whale. It is the only whale with asymmetrical colorization: the right side of its lower jaw is white, and the left side is pigmented. It can also move with unusual swiftness for a marine mammal. As with the blue whale, the fin whale inhabits every part of the world's oceans. Its largeness made it a favorite target of whalers, especially after blues became increasingly scarce, and its numbers plummeted between 1937 and 1965, when it was most intensely hunted. Nonetheless, the fin remains one of the most widely distributed and abundant of the great whales.[18]

Sei and Bryde's Whales

The sei whale and Bryde's whale closely resemble each other. The sei (*Balaenoptera borealis*) is so named because its annual spring migrations on the eastern Atlantic coincide with those of the *sei* (Norwegian for the fish variously known as saithe, pollack, and coalfish). Dark gray in color, the sei whale can attain a length of fifty to sixty-five feet and a weight of forty tons, making it only slightly smaller than the fin whale. Its annual migration is slightly less regular than that of other whales. Bryde's whale (*Balaenoptera edeni*)—forty-five feet in length and thirty tons in weight—is named after Johan Bryde, a Norwegian businessman who was one of the first to exploit these whales off the coast of South Africa. Its colloquial name, "tropical whale," is more informative. It roams in temperate and tropical waters, giving preference to the seas around Japan and South Africa; it does not migrate to polar waters. The sei whale and Bryde's whale were both considered too small to hunt in the early days of modern whaling, but they became prime targets between 1965 and 1975, when whalers could no longer find many blues and fins.

Minke Whales

The minke whale (*Balaenoptera acutorostrata*) is the smallest of the rorquals, reaching twenty-five to thirty feet in length and weighing around ten tons. The minke is black on top and white on bottom; it is the only rorqual that leaps completely out of the water when it breaches. It is distributed worldwide, but its southern and northern populations are quite distinct. Minkes were considered far too small to hunt as long as other rorquals were readily available, though they were targeted in the waning days of the whaling industry from 1975 forward. Still plentiful, they are among the only whales being commercially hunted today.

Biology was, for most whalers, an abstruse science about which they understood little or nothing. They did, nonetheless, possess an enormous amount of practical knowledge about cetacean anatomy and behavior that helped them track whales—including infants—with ruthless efficiency. They knew, for instance, that whales were easiest to spot and most vulnerable to a harpoon when they surfaced periodically to breathe or feed. They knew that whales followed predictable annual migration routes, that a group (pod) of whales was likely to assist a wounded member rather than flee danger, and that females almost never abandoned their young. They could recognize different whale species almost instantly—using size, color, blow patterns, and other distinguishing features as their guide—even if few of them could recite their Latin names. What they saw when they spotted a whale, however, was not a species per se but a floating warehouse—a cornucopia of oil, baleen, meat, bone, spermaceti, ambergris, and other valuable raw materials.[19]

Whale oil is the general term used to describe the edible oil extracted from the blubber, tongue, meat, bones, and internal organs of baleen whales. Structurally, it is a fatty acid, composed of one glycerine molecule combined with three fat molecules (triglycerides). Triglycerides are poor conductors of heat, making them ideal insulators in polar seas. The oil, once rendered and refined, can be used for illumination and lubrication. When combined with an alkali (such as potash or sodium carbonate), the fatty acids can be transformed into a good-quality soap. When mixed with vinegar, the oil can be applied to rice paddies as a pesticide. The glycerines in the oil can be used in the production of emollients and explosives. Most important, oleic acid, the most common fatty acid found in whale oil, can also be turned into margarine by adding two hydrogen molecules, a process known as hydrogenation. Until 1900, whale oil mostly went into the production of illuminants, soap, and explosives. By the 1930s, it was almost wholly used to make margarine and lard. (In 1934, some 84 percent of all whale oil shipped to Europe was turned into margarine.)[20] Whale oil is indistinguishable from other edible fats such as canola, soy, copra (coconut), and linseed except in one respect: it can be stored for a much longer period of time without spoiling (over five years). For that reason, many governments once maintained a strategic reserve of whale oil for use during wartime.[21]

Baleen is nicknamed "whalebone," but it is actually made of keratin, the same material found in fingernails and hair. Before the era of steel springs and plastic, it was a highly coveted product, often more profitable than whale oil. Its utility came from its lightness, flexibility, springiness, and strength. It was used in all kinds of consumer goods, including corset stays, hoop petticoats, umbrella ribs, ramrods, fishing rods, buggy whips, and carriage springs. It could be cut into strips and used as a sieve, a net, or a brush. When shredded, it could be used to stuff upholstered furniture. After steel and plastic came into widespread use in the late nineteenth century, baleen lost almost all of its commercial value and was discarded or used for mundane products such as chimney brooms.[22]

Whale meat has the look and texture of beef, but it is somewhat darker in color and has a stronger flavor. Only the Japanese and some indigenous peoples in the polar regions eat it on a regular basis. The Japanese distinguish between three types of whale meat: *akaniku* (red meat), which is similar in composition to that of land mammals; *onomi* (tail meat), which is especially fatty; and *sunoko* (ventral-groove meat), which is layered with connective tissue as well as fat. Medieval Europeans used to consume whale meat during Lent, treating it as fish instead of flesh, but modern Europeans and North Americans never developed a taste for it, except in whaling communities (such as Bergen, Norway), where it can be found on menus to this day. Since there was no market in Europe or the United States for the meat, whalers typically rendered it along with the bones and intestines in order to extract its oil; the waste meat was then used as animal feed, and the waste bone was turned into phosphoric fertilizer. The Japanese have been heavily criticized over the decades for using whale meat as a protein source, though it is not readily apparent why turning these animals into margarine and fertilizer is more justifiable.[23]

Sperm oil, spermaceti, and *ambergris* are unique to the sperm whale. Sperm oil and spermaceti are inedible waxes composed of esters of fatty acids with monohydric alcohols, making them structurally different from baleen oil (which contains glycerin). Sperm oil is rendered from the whale's body, and it is a liquid at room temperature. Heat-resistant and noncorrosive, it is an excellent lubricant for heavy machinery and can also be used for illumination. Spermaceti is found in a compartment (or case) in the whale's forehead. It is a liquid when inside the whale, but it turns into a milky white substance (hence the whale's sexualized name) when cooled and exposed to air. It is particularly suitable for the production of smokeless and odorless candles. Both sperm oil and spermaceti can also be used in the production of ointments, cosmetics, and soaps. The third product—ambergris—is a waxy material found in the large intestine or rectum of some sperm whales. It is probably an abnormal coalescence of substances that accumulates without doing the animal any apparent harm. Long a mainstay of the pharmaceutical, cosmetics, and alcohol industries, ambergris is still coveted today as a fixing agent for perfumes.[24]

The Impact of "Old Whaling" on Regional Populations

The Basques were the first people in the world known to have exploited whales primarily for their commercial value rather than for subsistence. Sometime in the twelfth century CE (and perhaps even earlier), they began plying the Atlantic seaboard between France and Spain in search of the right whale, whose migratory route included the Bay of Biscay, near the Basque homeland. Their killing technique—adopted by all subsequent

European whalers—was an adaptation of Stone Age technology to ocean conditions. A group of men rowed a boat as close as possible to a whale, and then one or more of them hurled a harpoon (the Basque word for spear) at its head or back. The harpoon was attached to the boat by a long rope. After spearing the animal, the boatmen's chief task was to stay alive while the whale exhausted itself dragging the vessel behind it, sometimes for days. Crew members then lanced the dying whale and towed the carcass to a shore station, where they flensed the blubber and rendered it in trypots until it melted into oil. Thousands of whales were caught this way, and many boats capsized, though the absence of written records makes it impossible to ascertain the total number.[25]

Over time, right whale populations in the Bay of Biscay began to dwindle, most likely because of overexploitation. By the thirteenth century, the Basques were sailing to the Arctic regions, especially to the waters around Greenland, where they found an untouched population of rights, bowheads, and Atlantic grays. By the sixteenth century, they had extended their hunting grounds to include Newfoundland and Labrador in eastern Canada, enlarging their annual take as they moved farther afield. They also began to use larger and more powerful vessels (including rowboats launched from sailing ships) and to flense the carcasses on the side of a boat, thereby reducing their dependency on shore stations. Their expeditions brought them into greater contact with northern Europeans—especially the Dutch, British, Danish, German, and Russian sailors along the North and Baltic seas—who began to imitate and eventually overtake them. The Dutch were particularly aggressive hunters, and by the seventeenth century, they had supplanted the Basques as the world's foremost whalers.[26]

The Dutch whaling industry was centered at first in Spitsbergen (also known as Svalbard), a cluster of islands east of Greenland. It was there that the Dutch established their first shore station in the 1620s—aptly named Smeerenburg (meaning "Blubbertown")—and began exploiting a largely untouched population of rights, bowheads, and grays. Like the Basques, however, the Dutch soon discovered that they had to keep moving farther afield as they depleted one hunting area after the next, and over time, they retraced the Basque trail to Greenland, Iceland, and eastern Canada. They hunted in much the same way as their predecessors, often using Basque harpooners and blubber-cutters on their expeditions, but their ships (especially the flyboat) were more versatile and numerous, and their collective impact was more lethal. Especially ominous was the total collapse of the Spitsbergen whaling grounds in the 1840s, after two centuries of exploitation, and the complete disappearance of the Atlantic gray populations (though no one knows for sure how or when that occurred).[27]

Among non-Europeans, the Japanese developed a unique and thriving whaling industry from the seventeenth century forward. The Basque method was unknown in Asia, but Japanese whalers developed their own hunting technique, one that resembles a cross between modern-day purse seining and samurai sword fighting. After a whale was spotted, boats surrounded it and directed it toward a series of large nets that were held fast between specially designed net boats. As soon as the whale was entangled in the nets, a group of men harpooned it until it was mortally wounded. Then several of them jumped on the whale's back and performed the coup de grâce with lance blows. This unique hunting method allowed the Japanese to take not only the easily catchable right whales (their favorite target) but also the more elusive humpbacks and even an occasional fin. (Pacific grays were taken too, but they broke through the nets and had to be harpooned.) From the outset, the Japanese hunted mostly for meat, not oil (Buddhist monks, forbidden to eat animal flesh, were especially fond of whale "fish"). They also took special care not to kill female right whales accompanied by young. This fact may well explain why the rights did not become endangered on the Japanese coastline as early as they did elsewhere.[28]

Whalers in the United States played a transitional role in the history of world whaling. For the most part, they used Basque and Dutch hunting techniques, but they also experimented with nearly all of the new techniques that would later be associated with modern Norwegian whaling. By and large, the industry was concentrated along the Yankee coastline of New England—New Bedford, Nantucket, Provincetown, New London, and Sag Harbor—though San Francisco also emerged as a major port during the nineteenth century. American-based enterprises emerged slowly during the seventeenth and eighteenth centuries, gradually expanding from the Atlantic to the Pacific, displacing the Dutch as they went. For a short period, from 1820 to 1860, the United States enjoyed a virtual stranglehold over the industry worldwide.[29]

Initially, Yankee whalers targeted the same whales as the Europeans—rights and bowheads—but by the early eighteenth century, they had become equally interested in sperm whales and grays. They developed the first safe and reliable method for rendering oil on the ship (a practice the Europeans had attempted unsuccessfully). Onboard tryworks turned U.S. ships into early versions of the floating factory, allowing whalers to travel farther from their base stations for longer periods of time, sometimes years. Then, in the mid-nineteenth century, Thomas Welcome Roys, the first whaler to exploit commercially the western Arctic region, invented two devices that foreshadowed the modern era. The first was a prototype winch-and-rope system (which Roys called a compensator but Foyn later renamed an accumulator) that reeled in a whale like a fishing rod. The second was a prototype bazooka gun, which allowed a gunner to shoot a harpoon from his shoulder.[30] ("It is now all over with the poor whales," Roys presciently remarked: "The weapon cleaves them like fate, making an internal wound about 10 feet in diameter closing at once every artery of life.")[31] So forward-looking were Roys's inventions that Norwegian historians have called him "the Svend Foyn of American whaling."[32]

Given Roys's inventiveness, U.S. whalers may well have nipped the Norwegian challenge in the bud, but domestic events conspired against them. In 1849, the California gold rush enticed thousands of able-bodied seamen into the Sierra Nevada, leaving many whaling expeditions bereft of experienced captains and crews. A decade later, much of the whaling fleet was sunk or scuttled during the Civil War (1861–65). Then, in the 1860s, kerosene—a cheap illuminant derived from coal and (later) petroleum—knocked the bottom out of the sperm-oil market. In 1846, some 735 whaling ships sailed under the U.S. flag, roughly three-quarters of the world's whaling fleet. The number had dwindled to 263 by 1866, to 124 by 1888, and to a mere 42 by 1906.[33]

The other great whaling nation of the premodern era was Great Britain. British whaling, however, did not follow the same pattern of peak and collapse that characterized the Basque, Dutch, and U.S. industries. Instead, it survived over many centuries, never becoming hegemonic but also never disappearing altogether. After going to war with the Dutch over whaling issues in the seventeenth century, the British settled for second best. This pattern repeated itself in North America, with British whalers first trying to control the U.S. whaling industry and then coming to accept it and even depend on it for oils. During the modern era, Britain once again played second fiddle, this time to Norway (even Britain's largest whaling enterprise, Christian Salvesen, was founded by a Norwegian immigrant). Like their competitors, British whalers concentrated their activities in the eastern Arctic, especially Spitsbergen, and then expanded westward until they reached Canada. From there, they ventured to the coastlines of Africa, South America, Australia, and New Zealand and finally to the heart of the Antarctic. British whalers would not pose a serious challenge to the Norwegians until the 1930s, but they did have two advantages that kept them competitive as others fell behind: many of the prime locations around the Antarctic perimeter were British colonies, and much of the world's whale-oil trade came to be concentrated in the hands of a British-based cartel, the Unilever (known as Lever Brothers before 1930).[34]

The Impact of Modern Whaling to 1931

Modern whaling began in the Varanger fjord of Finnmark, Norway's northernmost province, in the late 1860s. That is the fjord where Foyn first tested his new steamship and whaling harpoon on an untouched population of rorquals (mostly blues, fins, and seis) and also where he built his first shore station at Vadsø. Foyn's catch was modest by later standards—averaging only forty whales per year from 1873 through 1878 and ninety-two per year from 1878 to 1882—but these numbers were astonishing for their day because no one had ever managed to kill anywhere near that many rorquals before.[35] After a decade hunting in the Varanger fjord, Foyn and his growing band of competitors began to expand their operations, first to Iceland (around 1883), then to the Faroes (around 1894), to the Shetlands, the Hebrides, and Spitsbergen (around 1903), and finally to Ireland (1908), exploiting much the same eastern hunting grounds as their predecessors had. By 1894, they were collectively killing 1,528 whales, and by 1904 (the peak year), the number rose to 2,380 whales.[36] The total eastern Arctic tally for the entire period between 1868 and 1904 came to around 40,000 whales, almost all of which were rorquals and nearly all of which were killed by Norwegian gunners.[37]

The pickings were easy at first, much as in the early days of Basque, Dutch, and American whaling. But as competition grew and as whale populations declined in the eastern Arctic, companies found that they had to venture farther and farther afield. First, they went to Newfoundland and Labrador in eastern Canada (around 1898), then to British Columbia in western Canada (around 1905), and finally to Alaska (around 1912). Complete data for the North American catch are not available, but the Newfoundland catch stood at 858 whales in 1903 and 518 whales in 1909, and the Alaska catch stood at 482 whales in 1914 and 539 in 1919. The Japanese, meanwhile, adopted Norwegian techniques, and by the beginning of the twentieth century, they had transformed the coastlines of Japan and Korea into yet another rorqual killing site (with 42 killed in 1900 and 1,276 in 1920).[38]

For many decades, modern whaling was confined almost wholly to the Northern Hemisphere, close to known harbors and local markets. Only a handful of ships operated in the Southern Hemisphere at all, and those that did tended to be relics of former Dutch, British, and American enterprises. The geography of whaling, however, shifted dramatically when Carl Larsen, a Norwegian, captained the *Jason* and later the *Antarctica* deep into the southern seas in search of new killing grounds. Larsen went initially in pursuit of right whales, still the sentimental favorite among whalers, but he soon turned his guns on the rorquals—especially humpbacks, blues, and fins—that he found there in superabundance. Antarctica, however, posed unique challenges. First, it took an enormous amount of coal and time to reach the southern polar region from Norway. Second, the Antarctic summer was so short and bitterly cold that crews had to work at breakneck speed to fill their ships to capacity; all too often, carcasses froze solid before they could be flensed and processed. Third, ice hazards were everywhere, but protected harbors were few and far between. Given the hurdles, only the largest whaling companies risked their ships on such a costly experiment.[39]

Then, in 1904, Larsen found a solution: he established a shore station at Grytviken (meaning "Cauldron Bay"), an abandoned sealing site on South Georgia Island. South Georgia was part of an archipelago stretching from the Falklands to the Antarctic continent, under nominal British suzerainty. As soon as it proved a profitable spot for catching and processing whales, a wild "scramble for Antarctica" began. Rival companies built shore stations on South Shetland, South Orkney, and South Sandwich, all part of the Falklands archipelago. Others set up operations on South America's southern cone (Chile and Argentina), the African coastline (modern-day Mozambique, South Africa, Angola, Gabon, and Madagascar), or Australia and New Zealand—all locations that put them in striking distance of Antarctica.[40] Early on, the pickings were easy, much as they had once been in the Arctic. "We could often go right over to them and drop the harpoon in their backs," one Norwegian expedition leader noted: "It's unfortunate that cows with young should be killed, but there's nothing that can be done about it."[41]

By the 1909–10 season, Antarctic-based enterprises were already outpacing Arctic-based ones, both in terms of the annual catch (6,099 whales in the Antarctic as compared to 3,958 in the Arctic) and the amount of oil produced (157,592 barrels as compared to 112,347). From that moment on, "world whaling" was increasingly identified with the Southern Hemisphere in general and Antarctica in particular. By 1922, Antarctica's share of world whaling stood at 70.8 percent, and by 1930, it had reached 90.8 percent.[42] The number of whales taken each year also gradually increased. Between 1910 and 1924 alone, the Antarctic catch averaged around 9,000 annually (totaling 134,026 whales), enough to produce 433,000 barrels of oil per year

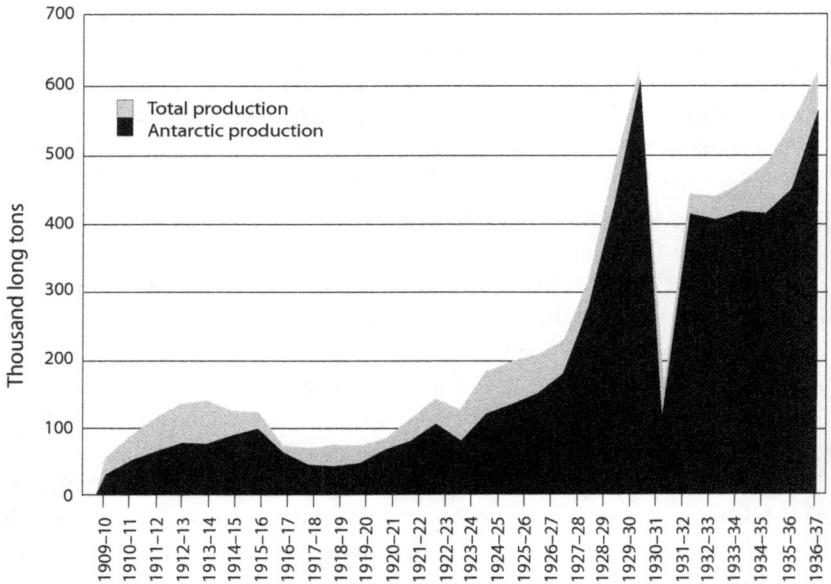

Figure 3.1. Global whale-oil production from the opening of the Antarctic whaling grounds in 1909 to the signing of the 1937 London Convention. Adapted from Committee for Whaling Statistics, *International Whaling Statistics*, vol. 16 (Oslo: Committeee for Whaling Statistics, 1930–88), 5.

(totaling 6,498,771 barrels).[43] These totals were nearly ten times the killing and production rates set by Foyn and his competitors during the heyday of eastern Arctic activities.

Then, in the 1920s, Norwegian whalers stumbled on the largest whaling grounds in the world: the Antarctic convergence (the meeting point of warm and cold ocean currents), where thousands and thousands of blues, fins, and other rorquals migrated each year to gorge on krill. Once again, it was Larsen who led the way when he ventured into the Ross Sea in 1923 aboard the vessel *Sir James Clark Ross*.[44] And once again, it was a Norwegian inventor, Petter Sørlle, who designed a new modern factory ship that allowed whalers to exploit these grounds. In 1925, Sørlle outfitted the *Lancing* with a stern slipway—a large trapdoor in the back of the ship that could be opened and closed as needed (much the way a car ferry works today)—as well as a ramp, winch, and whale claw. These tools enabled the crew to grab and hoist a whale into the main deck for flensing and processing before the carcass froze.

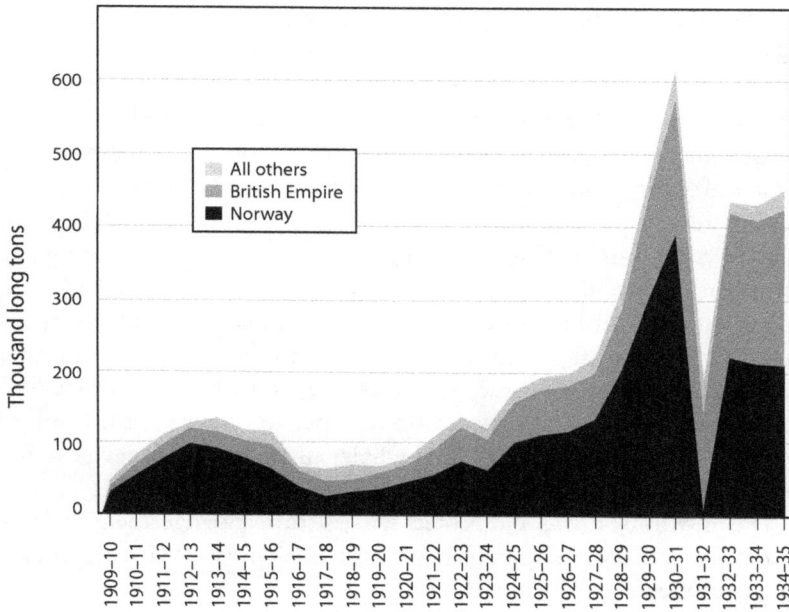

Figure 3.2. Global whale-oil production from 1909 to 1934 during the era of Norwegian and British domination. Adapted from Committee for Whaling Statistics, *International Whaling Statistics* (Oslo: Committeee for Whaling Statistics, 1930–88), vols. 1–12.

Sørlle's factory ship had another advantage: it freed whaling enterprises from their dependence on shore stations and thus ushered in the era known as pelagic whaling.[45] Whaling companies could simply anchor one or more of these factory ships in the middle of the krill grounds and use them as ersatz shore stations. Four or more whale catchers would accompany each factory ship to Antarctica and supply it with a steady stream of carcasses. By 1930–31, there were already 38 pelagic factory ships and 184 whale catchers operating in this manner, enough to triple the previous kill rate. The annual catch reached an all-time high of 47,200 whales (around 3.6 million barrels of oil) in 1930–31, then spiked again to 50,769 whales in 1937–38. Pelagic whaling also solidified the Antarctic's stronghold over world whaling. Back in 1868, nearly every rorqual death occurred at a single location—the Varanger fjord of northern Norway—in the Arctic. By 1938, 84 percent of the world catch occurred in the Southern Hemisphere, almost all of it from the flotillas operating in the krill grounds of the Antarctic convergence.[46]

With the advent of the modern Antarctic factory ship, the killing rate began to spin out of control. Whaling expeditions came to include not just a flotilla of factory ships, whale catchers, and fuel ships but also helicopters and other tracking devices (eventually including sonar). The demand for whale oil climbed steadily, as did (so it seemed) the opportunity for profits, and soon, there were new German, Japanese, Russian, and Dutch enterprises entering the fray. Indeed, by the 1930s, a "global commons" problem was beginning to emerge: all whaling enterprises understood that they had a collective interest in a sustainable annual harvest, but no individual whaling company had an incentive to take a unilateral step in the direction of reducing its annual kill rate. The expeditions were enormously expensive, and whalers felt they had to maximize their catch every season in order to turn a profit. Conservation on the part of an individual whaler just meant more whales for the others.

The only thing that could protect whales from endangerment and extinction—and the whaling industry from itself!—was governmental and intergovernmental regulation.

International Whaling Regulations to 1946

As with the African land animal treaties and the North American bird treaties, the whaling treaties emerged slowly and haphazardly over many decades. The early regulations were, for the most part, parochial in scope, designed to protect local or national interests rather than to safeguard the world's whale stocks. In 1863, the Norwegian government banned whaling in certain fjords during the herring season because local fishers were convinced (incorrectly) that whale carcasses attracted predators and scared off herring schools. In 1881, Norway banned all whaling within one mile of its coastline, and it imposed a total ban on whaling in the Varanger fjord during the cod season from January 1 to May 31. It subsequently banned whaling entirely in the territorial waters of Finnmark between 1904 and 1918, after a series of bad fishing years there. Similarly, in 1886, Iceland established a closed season on whaling within its territorial waters between May 1 and October 31, and it banned whaling entirely in the vicinity of herring fisheries. In 1915, it also imposed a ten-year ban on all whaling within its territorial waters (not to protect the whales but to keep the Norwegians out while it built a domestic whaling industry). Likewise, in 1902, the Danish government banned whaling around the Faroe Islands to all except those who flew the Danish flag.[47]

Asian whaling countries moved in the direction of regulation around the same time. In 1907, Korea established a closed season on whaling in its territorial waters between May 1 and September 30, and it placed a total ban on the killing of immature whales and adult females with young. These regulations were unusually forward-looking, for they focused on maintaining whale stocks rather than just national prerogatives. In 1908, the Japanese government channeled its whaling enterprises into a single umbrella organization, the Japanese Whaling and Fishing Association (headquartered in Osaka), and gave it the power to limit the number of whale catchers working in its territorial waters. This was forward-looking too in the sense that it predated the larger and more powerful Norwegian-led consortium of whaling enterprises, the Association of Whaling Companies (Hvalfangerforeningen), which was not founded until 1929.[48]

The British government, meanwhile, introduced two major innovations to whale management: (1) a fee-based licensing system, designed to discourage overfishing in any given hunting ground, and (2) a full-use requirement, designed to eliminate wastage and thus reduce the stench of rotting carcasses (one of the chief complaints of fishers). All whale boats operating in the Shetlands after 1906, for instance, had to pay an annual fee of £100 and agree to process their carcasses within sixty hours of capture. In eastern Canada (Newfoundland and Labrador), the right to maintain a shore station, along with a single whale catcher, cost $1,500 per annum as of 1902, and in 1928, the total number of shore stations, each now with two catchers, was capped at eight.[49] The French instituted a similar licensing system in their colonial dependencies and pushed for an international whaling treaty—as well as extending the coastal territorial zone to fifteen to twenty miles—though no progress in the direction of transnational cooperation came out of this initial effort.[50]

When whaling shifted from the Arctic to the Antarctic in 1904, Britain extended its licensing system there as well. The Falklands governor capped the total number of licenses for South Georgia at twenty-two. Each license, moreover, came with the right to establish only one shore station with a maximum of two whale catchers and one floating factory. In a glaring administrative oversight, however, he did not impose a full-use requirement until many years later, by which time thousands of discarded entrails and piles of meat were rotting on the island shores. Worse yet, he never imposed a yearly catch limit, thereby largely undermining the purpose of a licensing system. This situation resulted in the almost complete annihilation of humpbacks, the most prevalent species in the archipelago, within a short span of time. In the first decade of whaling operations, from 1904 to 1914, nearly 70 percent of the 29,016 whales killed in the South Georgian seas were humpbacks. This kill rate so vastly exceeded reproduction rates that a mere 131 humpbacks could be bagged during the 1917–18 season, prompting the governor to impose a temporary ban. Unfortunately, the ban was lifted long before humpback populations recovered, and from then on, they would never make up more than about 10 percent of the South Georgian catch.[51]

It would be too much to assert that the advent of pelagic whaling in the mid-1920s undermined these early attempts to regulate the whaling industry. The laws were too poorly constructed—and haphazardly enforced—to add up to anything remotely approaching an international regulatory regime in the first place. Pelagic whaling did, however, finally put an end to the *fiction* that a country-by-country approach could effectively regulate a global industry, and it raised the question of how (or if) the industry could be regulated in the future. Three institutions stepped in to fill the void. The first was the League of Nations (later the United Nations). League officials highlighted the importance of taking a biological approach to whale preservation, arguing that the needs of rational scientific management had to take precedence over the commercial rights of whaling nations. The League was the chief advocate of a comprehensive international treaty that would regulate the hunting grounds worldwide. The second association was the Association of Whaling Companies, founded in 1929 and known informally as the Sellers' Pool. Headquartered in Sandefjord, it had thirty-two charter members (twenty-five Norwegian companies, four British, two Danish, and one Argentinean), representing about 80 percent of world whale-oil production. Working closely with the Norwegian government, it was essentially a price-fixing cartel, its primary purpose being to keep the price of oil high enough to keep whaling profitable, even if it meant accepting an annual catch limit. The third association was the Unilever Group, a consortium of the three major margarine producers—Lever Brothers, De Nordiske Fabriker (De-No-Fa), and Margarine Unie—created in 1930 and informally known as the Buyers' Pool. Working closely with the British government, its primary purpose was to ensure that the price of whale oil remained at or below the price of palm, coconut, linseed, and other equivalent edible oils and fats. The various treaties, protocols, and bilateral agreements that emerged between 1931 and 1938 reflected the interplay of power and negotiation among these three institutions. Only the League of Nations took a scientific approach to whale management. The Sellers' Pool and the Buyers' Pool were solely interested in creating a reliable structure for producing and selling oil at a profit.

Conservationists had been discussing a whaling treaty at the same time they had been formulating the African and North American treaties. But the unregulated status of the world's oceans and the absence of precedents regarding the global commons hindered progress. It was only after the commencement of pelagic whaling in the mid-1920s that the International Council for the Exploration of the Sea (a League of Nations agency) began to explore the possibilities for a whaling convention. José León Suárez—an Argentinean diplomat and a whale-conservation enthusiast—was put in charge, and the committee's recommendations were published as the *Report on the Exploitation of the Products of the Sea* (hereafter the Suárez Report).

The Suárez Report made a number of remarkably forward-looking recommendations modeled in large part on the African mammal and North American bird conventions of the recent past. These included: (1) the creation of a rotation system for whale exploitation (similar to the three-field system in agriculture) in the Antarctic krill grounds, (2) an annual closed season during breeding times akin to those used to protect migratory land animals, (3) complete protection for all immature whales and their mothers, (4) standardization of capture methods, and (5) the implementation of a full-use requirement for all carcasses. "The riches of the sea," Suárez noted, "and especially the immense wealth of the Antarctic region, are the patrimony of the whole race."[52]

The Association of Whaling Companies saw the matter quite differently: whales were prey, not patrimony. At its urging, Norway's parliament passed the Norwegian Whaling Act of 1929 (hereafter the 1929 Norwegian Act), the chief purpose of which was to stave off a League convention or, failing that, to provide an alternative text to the Suárez Report for any future treaty. On the positive side, the 1929 Norwegian Act was far more comprehensive than any previous national law, and since Norwegian enterprises dominated world whaling, its impact extended well beyond the confines of Norway. Companies were forbidden to kill more whales than their floating factories could process before the carcasses began to rot. Waste was strictly forbidden: factory ships had to be outfitted with boilers and other equipment needed to render all parts of the whale (including the head, jaw, flank, tongue, and tail) into oil and to process other by-products, such as animal feed and fertilizer. The killing of right whales was forbidden outright, as was the killing of all calves, females with calves, blue whales under sixty feet long, and fin whales under fifty feet. To encourage the taking of mature whales, companies were required to pay their crews a wage based on the barrels of oil produced, not the number of whales caught. To ensure that these rules were enforced, each floating factory had to have at least one inspector aboard at all times. On the negative side, however, the law focused far more on the compilation of accurate records than on conservation. It did not limit the annual kill, the number of factory ships, or the yearly production. Nor did it establish a licensing system for pelagic whaling (though Norwegian companies were required to inform the government where they intended to send their floating factories before commencing operations).[53]

Once the 1929 Norwegian Act was in place, Norway and Britain used their influence over the League to ensure that its stipulations—not those of the Suárez Report—were used in the formulation of the Convention for the Regulation of Whaling, signed in Geneva in September 1931 (hereafter the 1931 Geneva Convention).[54] Missing from the 1931 Geneva Convention was any wording that highlighted rational management based on scientific principles; in its place was the raw language of commercial exploitation. Article 1 obligated the parties to "take appropriate measures" within "their respective jurisdictions" to ensure the "application of the provisions of the present Convention and the punishment of infractions." Article 2 made it clear that the convention applied only to "baleens or whalebone whales," not toothed whales. Article 4 stated in full: "The taking or killing of right whales, which shall be deemed to include North-Cape whales, Greenland whales, southern right whales, Pacific right whales and southern pigmy right whales, is prohibited." (In other words, all rights and bowheads received full protection worldwide.) Article 5 extended that prohibition to the "taking or killing of calves or suckling whales, immature whales, and female whales which are accompanied by calves (or suckling whales)."[55]

Article 6 required the "fullest possible use" of whale carcasses, stating in part: "Every factory, whether on shore or afloat, used for treating the carcasses of whales shall be equipped with adequate apparatus for the extraction of oil from the blubber, flesh and bones." Article 7 required companies to base the crew's pay (insofar as it was tied to production) primarily on "size, species, value and yield of oil taken" rather than on "the number of whales taken." This was supposed to encourage gunners to kill mature rather than juvenile whales, which it did. Unfortunately, it also encouraged gunners to target the largest of the largest whales, namely, pregnant blues. (In 1932–33, for instance, pregnant blues accounted for 80 percent of all whales over eighty-five feet that were killed that season.)[56] Article 9 noted that the treaty applied "to all the waters of the world, including both the high seas and territorial and national waters." Articles 10 through 12 dealt with the collection of statistics, and Articles 13 through 21 handled mundane issues relating to the process of implementation. Only two of these articles are noteworthy. Article 17 stated that the convention would enter into force only after it had been ratified by "eight Members of the League or non-member States, including the Kingdom of Norway and the United Kingdom." Article 19 limited the treaty's duration to three years after it came into force.[57]

The 1931 Geneva Convention deviated from the 1929 Norwegian Act in only two respects. Article 3 was added at the request of the Soviet Union. It excluded "aborigines dwelling on the coasts of the territories of the High Contracting Parties" from the terms of the treaty as long as they utilized "native craft propelled by oars and sails," hunted without the use of firearms, and did not work for or deliver whale products to commercial whalers. Article 8 was added by the British government. It required all commercial enterprises to have a valid license issued by the government under whose name its ships were registered. No limits were placed on the number of licenses each nation was allowed to issue, so in practice, it was ineffective as a conservation measure. At Canada's insistence, however, an additional clause was inserted at the end of Article 8: "Nothing in this Article shall prejudice the right of any High Contracting Party to require that, in addition, a license shall be required from his own authorities by every vessel desirous of using his territory or territorial waters for the purposes of taking, landing or treating whales, and such license may be refused or may be made subject to such conditions as may be deemed by such High Contracting Party to be necessary or desirable, whatever the nationality of the vessel may be." Countries, in other words, were free unilaterally to restrict the number of vessels in their territorial waters.[58]

Norwegian and British delegates fended off all efforts to further strengthen the treaty in the direction outlined in the Suárez Report. The Swedish government, for example, argued in vain for an additional clause that would have established a closed season. "In the Arctic Ocean, the whaling season begins at present on October 1st, when whales are comparatively thin and yield little oil," the Swedish delegate noted. "If the opening of the season were fixed, for instance, on December 1st, this would have two advantages: oil would be more easily obtained and fewer whales would be killed. In other parts of the world suitable close seasons might be fixed, based on biological study." Had the Swedish proposal passed, a precedent would have been set for the establishment of restrictive closed seasons and perhaps also the establishment of sanctuaries. Similarly, the Soviet Union unsuccessfully tried to add to Article 4 the following sentence: "It is absolutely forbidden to kill female cachalots [sperm whales] in any circumstances whatsoever." Had it passed, the convention's scope would have been widened to include the toothed whales. To their credit, however, Norwegian and British delegates also defeated all attempts to weaken the treaty's terms. Japan, for instance, tried in vain to exempt the North Pacific from the ban on hunting right whales. Likewise, Portugal failed in its attempt to exempt the waters around the Azores from the terms of the convention. And the Soviet Union was unable to insert a paragraph that would have obligated whaling companies to harpoon all killer whales they encountered (a clause that would have represented a throwback to the old days of so-called vermin eradication, had it passed).[59]

The 1931 Geneva Convention was signed in September and then sent to the various governments for ratification. The United States was the first to ratify, even though Article 6 (which imposed restrictions on manufacturing processes) and Article 7 (which guaranteed a minimum wage) were subject to constitutional challenge. Here, the 1916 Convention between Canada and the United States came to the rescue: the State Department was able to convince a wavering President Herbert Hoover to support the treaty because in *Missouri v. Holland,* the Supreme Court had upheld the power of the executive branch to "adopt methods which could not be valid if adopted by the legislative branch of the Government."[60] Once the United States ratified, so did Norway, the Union of South Africa, Switzerland, Mexico, the Netherlands, Italy, Spain, Poland, Czechoslovakia, Yugoslavia, Turkey, and Denmark. The British government, however, took three years to ratify, and as a result, the convention did not actually go into effect until the 1934–35 hunting season.

The League's secretary-general hailed the treaty as a great achievement that would "put an end to the uneconomic exploitation" of whales "without injuring the essential interests of the whaling industry."[61] Others, however, recognized that it was more a skeletal draft than a full-blown treaty and that the hard work of writing a genuine convention still lay in the future. It would be best, one British diplomat candidly noted, "to content ourselves at the moment with the present text of the draft Convention as laying down an elementary standard of conduct in whaling matters and providing a basis on which an effective system of control may gradually be built" and then "to utilize the period of the whaling holiday [the 1931–32 moratorium] by endeavouring to arrive at a closer understanding with the country chiefly interested, viz., Norway."[62]

At the same time, a temporary glut in the oil market acted as a momentary check on the whale slaughter, forcing producers to rethink their approach to whale management. Annual production yields had been zooming upward ever since pelagic whaling began. In the 1927–28 season, the total Antarctic catch stood at 13,775 whales (1,037,392 barrels), almost all of which came from 18 floating factories, 6 shore stations, and 84 whale catchers. By 1930–31, the total had climbed to 40,201 whales (3,608,348 barrels), from 41 floating factories, 6 shore stations, and 238 catchers. That translated into 611,014 metric tons of whale oil (two-thirds of which was produced by Norwegian expeditions)—the largest single-season production amount of all time.[63] The demand for oils and fats surged too, especially in Germany, but it did not keep pace with supply, especially after the onset of the Great Depression in 1929, and prices soon began to tumble. Oil fetched a price of around £30 per ton for most of the 1920s. Prearranged contracts kept the price propped up at £25 during the 1930–31 season, but it fell to £13 in 1931–32 and £11 in 1932–33. The outlook for profits was so bleak that the entire Norwegian fleet stayed in port for the 1931–32 season, as did most of the British fleet. Whaling resumed on a modest scale in the following year, but the industry did not begin to recover until the 1935–36 season, after weathering three rough seasons in a row.[64]

The Association of Whaling Companies (the Sellers' Pool) responded to the glut by signing a series of informal Production Agreements (June 1932, May 1933, and September 1936), all of which were negotiated completely outside the framework of the 1931 Geneva Convention. These side agreements assigned each of its members a catch quota based on past production rates, and they capped the total annual oil yield at 2.2 million barrels. The sole purpose of these side agreements was to raise the price of oil artificially by keeping production about one-third below the 1930–31 season. Rarely was price-fixing accomplished in a more open or blatant form.

To verify compliance, the association concocted the Blue Whale Unit (BWU), a notorious conversion standard that would remain in use until it was finally banned by international treaty in 1974. The BWU was based on the fact that the average blue whale produced 110 barrels (4,400 gallons) of oil, roughly twice as much oil as a fin whale, two and a half times as much as a humpback, and six times more than a sei whale. The BWU conversion formula was thus 1:2:2.5:6. The beauty of the system, from the whalers' vantage point, was that a catch quota could be set each year without reference to the whale species. If the total quota for any given year was set at 20,000 BWU, then whalers were free to kill 20,000 blue whales, or 40,000 fins, or 50,000 humpbacks, or 120,000 seis, or any combination thereof that totaled 20,000 BWU.[65]

The problem with the BWU system, from the nonwhalers' vantage point, was that it worked against whale conservation. First, the annual BWU quotas were keyed to the market price of oil, not to the reproduction rates of whales. Second, the stock of whale species in the world's oceans did not conform to the 1:2:2.5:6 formula; there were not 2 fins for every blue in the world's oceans, any more than there were 2.5 humpbacks or 6 seis for every blue. Third, the BWU put a premium on size, since a company could fill its quota with less effort by taking the largest species. The BWU system thus promoted maximum profitability but not maximum sustainability.

Predictably, from that moment on, whaling enterprises targeted blues first, fins next, humpbacks next, and seis last, moving down the size chart as each of the larger species became too rare to hunt commercially. All that the BWU system really did was to put a bull's-eye on the largest available species until it was no longer plentiful, ultimately a self-defeating strategy. Even for the whaling industry itself, the BWU system only worked properly in conjunction with a yearly side agreement that imposed enterprise-by-enterprise production quotas. When these side agreements ended in 1936, enterprises were left with a strong incentive to construct larger factory ships and faster catcher boats in order to capture the largest possible share of the total annual catch before the BWU limit was reached. This in turn made it all the more difficult to adjust the BWU limits downward in light of conservation needs: a high annual quota was necessary to guarantee that whaling companies would recoup the huge investments they made in their flotillas. In a nutshell, the BWU system promoted an "arms race" rather than a conservationist ethic among the whaling companies.[66]

While the Association of Whaling Companies was busy establishing the BWU system to prop up the price of oil, the Unilever Group (Buyers' Pool) was doing what it could to drive prices downward. Unilever was a tight-knit consortium of margarine producers with a major trump card: it could switch to coconut, palm, peanut, or soybean oil if the price of whale oil rose beyond what it considered to be an acceptable price. Unilever's position was further strengthened by its connections to Christian Salvesen, Britain's largest whaling enterprise. Not only was Salvesen one of the few whaling enterprises that did not belong to the Association of Whaling Companies, it also actively worked to undermine the Production Agreements (for instance, it sent its fleet to the Antarctic during the 1931–32 season while the entire Norwegian fleet stayed home and then used its catch from that year to flood the market whenever the price began to climb). With plenty of other edible oils available and with Salvesen acting as a strategic reserve, Unilever strove to keep the price under £14 per ton. This price was well below the £32 per ton that the whaling enterprises had enjoyed during the previous five seasons, and beyond that, it was at or below the cost of production, which (depending on enterprise) ranged from £14 to £18 per ton.[67]

The tug-of-war between the Sellers' Pool and the Buyers' Pool, of course, had nothing to do with whale protection or catch management and everything to do with the protection of narrow economic interests. In a free market, the price may well have fallen below the cost of production, and many whaling companies have gone belly up before the price rose anew. More plant-based oils would have reached consumers, and more whales would have remained in the ocean to reproduce. However, the two whaling pools (and the two governments, Norway and Britain, that backed them) conspired to keep the industry sheltered from market-based competition, each for its own purposes. The Sellers' Pool wanted to keep its operations going full tilt for as long as possible each season, since ships cost money to maintain regardless of whether they were out to sea or not. The Buyers' Pool wanted an array of different edible oils at their disposal, so as to pit sellers against each other and thus keep all oils at an acceptable price. Neither pool was interested in conservation, a point that Birger Bergersen, a Norwegian biologist and later the first chair of the International Whaling Commission, hammered home at a whaling conference in May 1938: "In years when the prices are low and when we know that whale oil can be substituted by vegetable oils, it seems to the biologists that it is too bad that so many whales should be killed. It is easy to get plenty of new trees in the tropical areas, but it is impossible to renew the stock of whales when it is overtaxed."[68]

Norwegian and British authorities soon began to see the flaws in the 1931 Geneva Convention and the informal Production Agreements, and they formulated a new set of national regulations and bilateral agreements. Two stand out. In 1935, Norway amended its 1929 Whaling Act so that it reduced the Antarctic season to the period between December 1 and March 15, restricted the hunting grounds to south of 40° south (deep in the heart of the Antarctic), and capped the total production by Norwegian companies at 1.1 million barrels. This signaled Norway's willingness to accept short seasons and large ocean sanctuaries—but not a limit on the number of whales that could be caught each year or the number of factories that could operate in any given season. In 1936, the Norwegian and British governments jointly truncated the season for blues and fins (from December 8 to March 7), capped the number of catcher boats at seven per factory ship, and introduced production quotas.[69]

Unfortunately, at the very moment when the Norwegian and British governments were finally beginning to rein in their own enterprises in the interest of genuine conservation, they began to lose their near monopoly on the global whaling industry. The Japanese joined the Antarctic pelagic

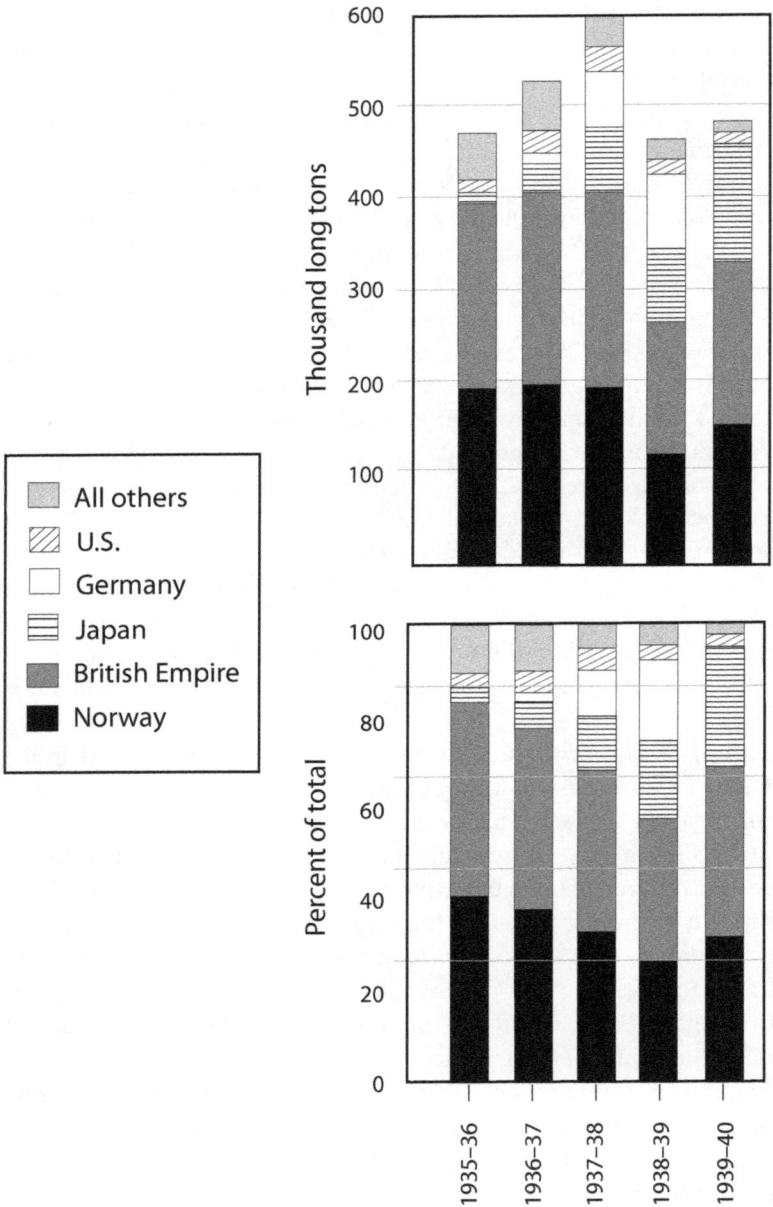

Figure 3.3. Global whale-oil production from 1936 to 1940 after the breakdown of Norwegian-British domination. Adapted from Committee for Whaling Statistics, *International Whaling Statistics* (Oslo: Committeee for Whaling Statistics, 1930–88), vols. 1–12.

club in 1934–35 with the purchase of the aging factory ship *Antarctica* (renamed *Tonan Maru*, or Southward Aspiration) from a Norwegian firm. By the 1938–39 season, Japan had six factory ships operating in the Antarctic, enough to capture about one-sixth (17.1 percent) of the world's whale-oil market. Whaling also fit into Nazi Germany's "Fat Plan," an effort to attain autarchy in the production of edible oils, announced shortly after Adolf Hitler came to power in 1933. In 1937–38, the Germans launched their first whaling ship, the *Walter Rau,* named after the country's chief margarine magnate and whaling enthusiast. Germany's ambitions came to an abrupt end with the outbreak of World War II, but for a brief moment, the country produced about one-eighth of the world's whale oil (peaking at 13.2 percent in 1938–39). More to the point, Japan and Germany collectively accounted for 84 percent of the production *increase* that occurred between the 1936–37 and 1937–38 seasons. Moreover, neither country had signed the 1931 Geneva Convention, and they were therefore not bound by the same hunting restrictions as their competitors.[70] (The Soviet Union also emerged as a whaling power in the 1930s but did not begin Antarctic pelagic whaling until 1946–47, after seizing one of Germany's factory ships.)

Once they realizing that bilateral negotiations no longer worked, Norway and Britain decided to update and renew the 1931 Geneva Convention (which was set to expire at the end of the 1936–37 season), preferably with Germany and Japan as participants this time. The British agreed to host the conference and to draft the International Agreement for the Regulation of Whaling (hereafter the 1937 London Agreement), which was subsequently signed and ratified with few alterations to the draft text.[71] By and large, the 1937 London Agreement was an amalgam of the 1931 Geneva Convention and the Production Agreements. The preamble laid out two potentially contradictory goals: "to secure the prosperity of the whaling industry" and "to maintain the stock of whales." Articles 1 through 3 made it mandatory for all factory ships to have at least one government inspector on board at all times and outlined the methods of enforcement. Article 4 reiterated the ban on hunting right whales and extended protection to grays for the first time. Articles 5 and 6 established minimum lengths for the taking of blues (70 feet), fins (55 feet), humpbacks (35 feet), and sperms (35 feet) and reiterated the ban on killing mothers and calves.[72] These lengths were adopted even though Bergersen, Norway's delegate, pointed out that scientists had recently established that the minimum length for sexual maturity in female blues was 78 feet and in female fins 66 feet.[73]

Article 7 established a closed season for pelagic whaling in the Antarctic region, stating in part: "It is forbidden to use a factory ship or a whale catcher attached thereto for the purpose of taking or treating baleen whales in any waters south of 40° South Latitude, except during the period from the 8th day of December to the 7th day of March." Article 8 imposed a different set of restrictions on shore-based operations: "It is forbidden to use a land station or a whale catcher attached thereto for the purpose of taking or treating whales in any area or in any waters for more than six months in any period of twelve, such period of six months to be continuous." Article 9 forbade the use of factory ships and whale catchers to capture baleens in four geographic regions: (1) in the Atlantic Ocean, north of 40° south latitude and in the Davis Strait, Baffin Bay, and Greenland Sea; (2) in the Pacific Ocean east of 150° west longitude between 40° south latitude and 35° north latitude; (3) in the Pacific Ocean west of 150° west longitude between 40° south latitude and 20° north latitude; and (4) in the Indian Ocean north of 40° south latitude. This article was more symbolic than anything else, since almost all pelagic whaling took place in the two areas not covered by the treaty—the Antarctic and the North Pacific. Finally, Article 10 allowed each national government to take as many whales as it "thinks fit" for "purposes of scientific research."[74]

The remaining articles followed the 1931 Geneva Convention guidelines with only minor language changes, except in two regards. Article 12 made it mandatory that all whales be processed within thirty-six hours of being killed to ensure that the oil was of the highest quality. Article 18 listed in great detail what was meant by the terms *factory ship, whale catcher, land station,* and *baleen whale.* It also identified the *blue whale, fin whale, gray whale, humpback whale, right whale,* and *sperm whale,* along with their common nicknames, for greater clarity and precision.[75]

Norway, Great Britain, Argentina, Australia, Germany, Ireland, New Zealand, South Africa, and the United States all signed the 1937 London Agreement in time for the 1937–38 hunting season. These same countries then met again the following year in London, along with Canada, Denmark, France, and Japan, to sign the Protocol Amending the International Agreement of June 8, 1937 for the Regulation of Whaling (hereafter the 1938 London Protocol). The main purpose of this follow-up meeting was to take up some issues that had proven too contentious in 1937. Attitudes, however, had not softened, so there was little improvement and indeed some backtracking. A majority of the delegates still rejected a shortened hunting season. They still refused to limit the number of catchers that could operate in conjunction with a single factory ship or land station. They still would not accept a quota for each factory ship. They still would not cap the total Antarctic catch. They voted down a plan to divide the

Antarctic into distinct hunting areas (later established as Areas I through VI), a major prerequisite for effective protection of still-untouched portions of the Antarctic. They even refused to give full protection to humpbacks, despite the growing evidence that this species was as endangered as rights and grays, deciding instead to prohibit humpback hunting south of 40° south latitude for the 1938–39 season only. Worse yet, as a concession to Japan, they reduced the minimum lengths for blues, fins, and sperms by five feet for any whales delivered to land stations "for local consumption" as human or animal food.[76] "It might have been nice to return home with a little bit more in the way of results," Bergersen later admitted.[77]

Progress was made in only one area. Article 2 of the 1938 London Protocols established, for the first time, a whale sanctuary in the western Antarctic (south of 40° south latitude from 70° west longitude westward as far as 160° west longitude) for a period of two years beginning December 1938. This sanctuary lay in the Southern Ocean between Cape Horn and the Ross Sea, in a part of the ocean that biologists knew contained numerous cetacean species but where commercial whaling had not yet commenced. Here, the migratory bird treaties between the United States, Canada, and Mexico served as a model. "The North American populations of migratory waterfowl have been restored from the low point of 27 million individuals [in] about 1930 to a total of 125 million in 1945 in part by the operation of sanctuary areas," a U.S. whaling negotiator noted in defense of whaling sanctuaries. "Without the sanctuaries it would have been very difficult, if not impossible, to protect adequate breeding stocks (Treaties between U.S.A. and Canada, and U.S.A. and Mexico)."[78]

The major whaling powers met one more time in London (in July 1939), at which point they further watered down the regulations as a sop to Japan, but World War II began shortly thereafter, bringing an end to the annual meetings and to the 1937 London Agreement. Its usefulness as a conservation treaty was, in any case, severely circumscribed. During the 1937–38 season, a record number of whales were legally killed— 50,769—the largest catch ever. This fact reflected the diplomats' failure to establish an annual production cap or to limit the number of factory ships and whale catchers operating in the Antarctic. Meanwhile, whalers took record numbers of fins, a clear sign that blue populations had become depleted after decades of overfishing and that gunners were now targeting the next-largest species for decimation. Finally, Japan (which never ratified any of the London accords) allowed four of its expeditions to stay at sea for 125 days, nearly a month longer than the agreed-upon hunting season of 98 days. Without a catch limit, a cap on whale catchers, or Japanese participation, the 1937 London Agreement and its follow-up protocols were doomed to failure.[79]

None of these obvious problems with maintaining the whale stocks over the long haul, however, ever seemed to penetrate into the inner circles of the whaling industry or dampen the enthusiasm of global investors. "Whaling is no longer the romantic and haphazard affair beloved of the writers of adventure stories," *Financial News* confidently proclaimed in December 1937: "It is now rationalized and highly mechanized, has boards of directors and shareholders, close seasons to regulate the slaughter of its victims and floating commissions to enquire into their habits and ways of living. In short, the whale is now killed according to the rules of science."[80]

Whales received a respite during World War II (a war in which millions of humans were also "killed according to the rules of science"), though the hunting hiatus was not long enough to replenish the global stocks to any substantial degree. Some whaling continued to take place during the war years, but yearly production rates fell to levels not seen since the first decade of the twentieth century, when Antarctic whaling was still in its infancy. Whale diplomacy also came to a near standstill. Britain, the United States, and Norway met only once during the war, in January 1944, ostensibly to address the looming fat shortage in Europe but in reality to plan for the resumption of whaling after the war. They decided to extend the hunting season to four months (November 24 to March 24), to establish an annual quota of 16,000 BWU, and to compose a new convention in 1945 (later delayed to 1946). Unfortunately, the extended season worked against conservation without resolving the fat shortage: whales contain about 20 percent less blubber in November than they do in February, after they have gorged themselves on krill. It would have been wiser from an economic standpoint as well as a conservationist one to have stayed with a later starting date. The new quota figure, moreover, was based on speculation rather than science, for no one knew how many whales were still in the oceans, where they were concentrated, or how many could be killed each year on a sustainable basis. The three scientists who picked the figure—Bergersen (Norway), Remington Kellogg (United States), and N. A. Mackintosh (Great Britain)—simply took the production rates for the two seasons prior to the outbreak of World War II (29,876 BWU and 24,830 BWU) and reduced them by roughly 40 percent. Despite the guesswork, the 16,000 BWU figure was, from that moment on, wrapped in an aura of scientific soundness, and it remained the norm until the early 1960s, by which time the whale stocks were so depleted that the quota could not even be filled.[81]

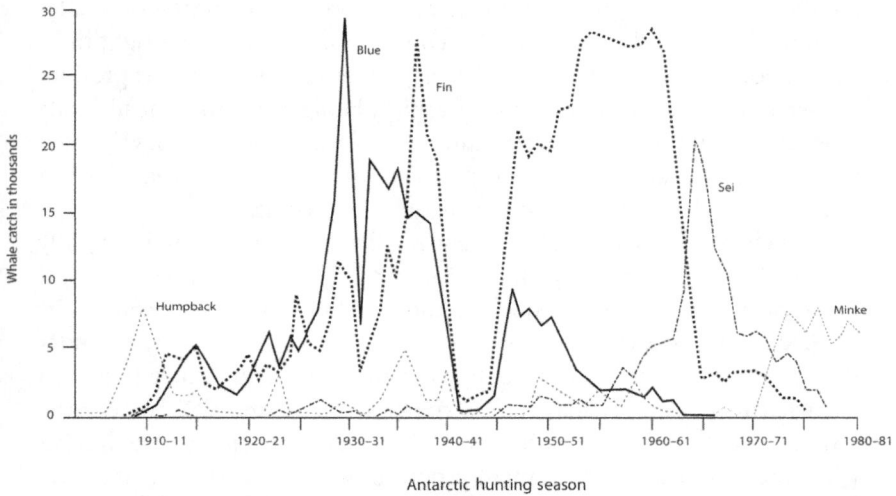

Figure 3.4. Whale catches in the Antarctic region since 1900. Note that from the 1920s onward, whalers targeted the largest readily available species first, starting with the gigantic blues, then the fins, and finally the seis, depleting each species in turn. The whaling business was stopped in the 1980s before the minkes, the smallest of the great whales, had been devastated. Adapted from Peter G. H. Evans, *The Natural History of Whales and Dolphins* (New York: Facts on File Publications, 1987), 257.

At the end of World War II, the major whaling powers met once again to negotiate the International Convention for the Regulation of Whaling (hereafter the 1946 Washington Convention), which they signed in Washington, DC, on December 2, 1946. The United States organized the conference, composed the first draft (hereafter the U.S. Draft), and dominated the negotiations. Only a handful of changes were made to the U.S. Draft; almost all of its most important features were accepted.[82]

The preamble offered a candid rationale for an international treaty: "to provide for the proper conservation of whale stocks and thus make possible the orderly development of the whaling industry." It optimistically noted that "whale stocks are susceptible of natural increases if whaling is properly regulated, and that increases in the size of whale stocks will permit increases in the number of whales which may be captured without endangering these natural resources." It recognized that "whaling operations should be confined to those species best able to sustain exploitation in order to give an interval for recovery to certain species of whales now depleted in numbers." And it admitted that "the history of whaling has seen overfishing of one area after another and of one species of whale after another to such a degree that it essential to protect all species of whales from further overfishing."[83] The U.S. Draft was even more forthright in its declarations—"many whale fisheries may never recover," "the few productive whaling areas now remaining are rapidly being depleted," "the ultimate objective should be to achieve and to maintain the stocks at a level which will permit a sustained capture of the maximum number of whales"—but none of these phrases survived the negotiating process.[84]

Much of the 1946 treaty merely reaffirmed (sometimes in slightly modified form) the verbiage that appeared in previous treaties. Article I made clear that the convention (and the schedule that was attached to it) applied to all "factory ships," "land stations," and "whale catchers" under the jurisdiction of the "Contracting Governments," and Article II spelled out what was meant by these terms. Articles IV, VII, and VIII laid out the ground rules for scientific research and data collection, and Article IX dealt with labor issues and enforcement procedures. Article X stipulated that at least six countries had to ratify the treaty before it went into effect and that Norway, the United Kingdom, the Netherlands, and the USSR had to be among those that ratified. It also stated that "any Government which has not signed this Convention may adhere thereto after it enters into force by a notification in writing to the Government of the United States." (This mundane-sounding declaration would take on importance decades later when nonwhaling nations ratified the convention for the sole purpose of banning whaling.)[85]

For the most part, the schedule attached to convention simply reiterated the conservation measures of earlier treaties and agreements on closed seasons, sanctuaries, minimum sizes, and protection for endangered and immature whales. There were, however, some notable changes. Each ship now had to maintain two inspectors, not one. Factory ships that operated exclusively within the territorial jurisdiction of certain African and Australian regions were to be treated as if they were land stations. The BWU formula (1 blue whale equals 2 fins, 2.5 humpbacks, or 6 seis)—which had been an informal part of all whaling negotiations since 1931—was now formally adopted. The total annual yield was set at 16,000 BWUs, the same number that had been in effect (but never reached) during the war.[86]

Articles III and V offered innovations not seen in previous whaling treaties. Article III established the International Whaling Commission (IWC), which remains the most important agency regulating whale hunting today. Each country that ratified the treaty was allowed to have just one member on the IWC and also just one vote. The IWC was allowed to rely on a simple majority for making most of its decisions, but changes to Article V required the consent of three-quarters of the members (somewhat higher than the two-thirds majority called for in the U.S. Draft). Article V allowed the IWC to "amend from time to time the provisions of the Schedule by adopting regulations with respect to the conservation and utilization of whale resources." This provision gave the IWC the right to protect endangered species, establish closed seasons, designate sanctuaries, establish size limits, cap the annual catch, and ban hunting techniques without the need to secure a new treaty.[87] Originally, the United States wanted to place control over whaling in the hands of the UN's Food and Agriculture Organisation, but Norway successfully championed a freestanding commission outside the scope of the United Nations.

After agreeing to the creation of the IWC, the contracting governments made sure that there were some strong checks on its regulatory power. Article V prohibited the commissioners from placing "restrictions on the number or nationality of factory ships or land stations, nor allocate specific quotas to any factory ship or land station or to any group of factory ships or land stations." This prohibition was designed to make it impossible for the IWC to favor the enterprises of one whaling nation over those of another, but it also made it impossible to halt the proliferation of whaling nations and enterprises. Article V, moreover, also allowed the contracting governments to exempt themselves from the terms of any IWC amendment simply by lodging an objection to it within ninety days. This measure effectively gave each whaling nation a unilateral veto over important IWC decisions, since no nation could be expected to adhere to a restriction when a rival nation did not. Finally, Article XI allowed a government to withdraw from the convention for the next hunting season simply by announcing its attention on or before January 1.[88] Whaling nations used these powers to undermine the IWC's effectiveness on numerous occasions.

The British delegation wanted to create an Article XII with two passages. The first would have required the contracting governments to "take all practicable steps to prohibit the sale, charter, transfer, loan or delivery of vessels, equipment or supplies designed especially for whaling operations or known to be intended for such operations, to any Government or any person operating under the jurisdiction of any Government not a party to the Convention." The second would have required them to "take all practicable steps to prohibit the import into territory under the jurisdiction of any of the Contracting Governments of whale oil or other products produced under the flag of any non-Contracting Government." These additions would

have strengthen the treaty's ability to thwart whaling by nonparty countries and pirate whaling by private persons. The U.S. delegation, however,
successfully prevented the inclusion of Article XII on the grounds that it
"would constitute a violation by the United States of trade agreements to
which the United States is a party" and because "coercive economic measures are not appropriate in connection with a long-range conservation
agreement."[89]

The 1946 Washington Convention was the single most important cetacean
treaty of the twentieth century. It went into effect for the 1948–49 season,
and it has governed whaling affairs ever since. Unfortunately, it did little
more than provide a legal framework for continuing the unsustainable exploitation of whales.

Article V contained the largest number of stumbling blocks to conservation. It prohibited any restrictions on the number or nationality of
factory ships and land stations and made it impossible to introduce ship-
by-ship or country-by-country quotas. These prohibitions had the effect
of encouraging the proliferation of whaling nations at a time when the
stocks were already dangerously depleted. At the end of World War II, only
Norway and Britain were still in a position to engage in pelagic whaling,
but within a few years, the Soviet Union, Japan, and the Netherlands had
joined (or rejoined) them. During the 1946–47 season, there were just fifteen expeditions using 129 catchers, but by 1961, there were twenty-one expeditions using 261 catchers.[90] Since the annual allowable catch remained
steady at 16,000 BWU, an ever-increasing number of enterprises had to
compete for the same amount of oil.

Article V's noncompliance clause also contributed to the decline in
whale stocks. In 1954, for instance, Canada, the United States, Japan, and
the USSR objected when the IWC imposed a ban on the hunting of blue
whales in the North Pacific. In 1981, Brazil, Iceland, Japan, Norway, and the
USSR objected to the ban on the use of the "cold grenade," a cruel killing
tool used on minkes. And in 1982, Peru and Chile objected to IWC restrictions on the killing of Bryde's whales. In each case, the objections rendered
the IWC's efforts at regulation totally useless.[91]

Cutthroat competition on the part of the whaling companies, together
with a devil-may-care attitude among the whaling nations, made it politically impossible for the IWC to reduce the 16,000 BWU quota in the interest of conservation: a smaller quota would translate into bankruptcy for
one or more whaling enterprises, a situation that no government wanted
to face. "The system of free competition for a fixed overall catch had almost

all the defects of a competitive system, with none of the advantages," G. H. Elliot noted in his postmortem analysis of the IWC: "Each operator was fully exposed to the actions of its competitors; each invested more and more to obtain a larger share. The investments cancelled each other out, and resulted only in a shorter season, in which efficiency of processing and the development of by-products was sacrificed to catching and working up as many whales as possible before the catch limit was reached."[92]

Enforcement presented an equally thorny problem. The treaty left it up to each member nation to ensure that the whaling firms flying under its flag followed the regulations. From 1946 to 1963, however, the USSR openly flouted the rules, taking whales of all sizes and species in and out of season. Soviet authorities finally agreed to accept international observers on their factory ships in 1963 but then failed to implement this arrangement until 1972, by which time many whale species were already endangered. Even then, they worked out an arrangement whereby Japanese observers would be aboard Soviet ships and Soviet observers aboard Japanese ships, each side winking when the rules were violated.[93] Pirate whaling exacerbated the problem. Between 1949 and 1956, for instance, Aristotle Onassis operated a factory ship, the *Olympic Challenger*, year-round, as did other less flamboyant entrepreneurs, usually flying the flag of Panama or some other country that was not party to the 1946 convention.[94] The track records of Britain, Norway, and the Netherlands were much better, but it was an open secret that factory inspectors often turned a blind eye when gunners inadvertently killed an immature or protected whale.

Without effective enforcement mechanisms, the careful work of IWC scientists—organized first as the Scientific Committee and later as the Committee of Four—bore almost no positive results. In 1955, for instance, the Scientific Committee requested that the annual BWU be reduced each year by 500 BWU until it reached 12,000 BWU (itself an overly optimistic figure) in order to guarantee sustainable yields. This recommendation, however, got nowhere due to the objections of the Netherlands, and as a result, the 16,000 BWU quota remained largely intact even as the stocks dwindled to the point that this quota could not even be filled. Then, in 1960, the Committee of Four recommended that the BWU system be abandoned entirely in favor of a species-by-species approach to whale management, and it argued for a total ban on blue and humpback hunting as well as an annual cap on fins (nearly twenty-eight thousand fins were killed during the 1960–61 season alone). The IWC used this moment to reduce drastically the annual catch to 2,700 BWU but not to introduce a more sensible, species-by-species approach.[95] "Almost all major actions or failures to act," the IWC commissioner for the United States, J. L. McHugh, later noted, "were governed by short-range economic considerations rather than by the requirements of conservation."[96] A member of the IWC's Scientific Committee, Sidney Holt, was even more blunt: "Whaling is essentially an extractive industry, akin to mining. Targeted depletion of one whale 'seam' stops when it becomes uneconomic to extract more, and the industry moves on to other places and species."[97]

That the 1946 Washington Convention did nothing to halt the killing spree can be seen from the official whaling statistics. Between 1905 and 1965, roughly 1.25 million whales were slain in the Antarctic. Just over half of these were killed in the forty-three-year period between 1905 and 1948. The remainder were killed in the seventeen-year period between 1948 and 1965, when the treaty was in effect and the IWC operational.[98] By the mid-1960s, whaling companies had managed to deplete most of the remaining stocks of great whales almost as quickly as they would have without a treaty. Blue whales stood on the brink of extinction. Fin populations plummeted to record lows, and Bryde's and seis were endangered. Only the minke, the smallest rorqual, remained plentiful worldwide. As the stocks dwindled, so did the number of pelagic nations. Great Britain largely abandoned whaling in 1963, the Netherlands in 1964, and Norway in 1968. Others followed suit, and by the 1970s, there were only two major players left in the field—the USSR and Japan—and they were increasingly forced to hunt the minke, once considered too small to be worth the chase.

The 1946 Washington Convention was inadequate as a hunting treaty, let alone as a conservation treaty. Hunting treaties aim to keep the stock of coveted species at the highest possible levels in order to maximize the hunters' enjoyment. Game reserves, hunting seasons, protection for females and juveniles, and similar restrictions are among the most effective techniques for accomplishing this goal. Conservation treaties aim to keep the stock of all species, whether hunted or not, at the highest possible level in order to maintain a more balanced (and therefore less volatile) ecological system. The 1946 Washington Convention had the opposite impact: it promoted the decimation of one hunted species after the next, moving down the line in size, leaving only minke populations intact. The IWC did regularly step in to protect species once they had become too rare to hunt profitably, and these efforts certainly helped protect remnant whale populations from complete extinction. But a treaty that protects species only after they have been hunted to the brink of commercial extinction can hardly be held up as an example of sound international cooperation.

The 1946 Washington Convention failed at a number of levels. It did not provide adequate protection to females and calves during the breeding season; it did not set minimum size limits at the optimum level; it did not regulate the open season properly; it did not limit the number of hunters; and it did not establish adequate sanctuaries. These failures were in turn largely rooted in the fact that there was no international agency that could control the open seas to the same degree that the countries involved in the African and North American treaties could control their land and skies. Whales would have fared much better if the League of Nations had implemented the Suárez Report, with all the safeguards that it envisaged. Whales would have probably fared much better if Great Britain or the United States (or even Norway) had controlled the Antarctic region because decades of domestic debate over a variety of conservation issues had made them attuned to the need for whale conservation. The Antarctic convergence, however, did not belong to the League any more than it belonged to any individual country, and therefore, all attempts—meager as they were—to introduce conservation measures only encouraged proliferation and evasion.

The whaling companies, not the conservationists, held all the trump cards in the open seas, and they were more interested in today's profits than tomorrow's bounty. The treaties they helped craft and the commission they helped establish were therefore designed more to protect their industry than to protect the world's stock of whales. "The Whaling Commission," polar biologist Bernard Stonehouse later observed, "faced the impossible task of controlling a powerful, profitable, highly capitalized, fiercely competitive, multinational industry—one which had no intention of accepting controls other than on its own terms."[99]

Ironically, the whaling enterprises doomed themselves even as they passed a death sentence on the whales, for their continued existence depended wholly on the existence of a sustainable harvest. A mere hundred years after Foyn invented the exploding harpoon, commercial hunters had come perilously close to killing off the largest mammal ever known to exist. The situation was eerily presaged by Herman Melville back in 1851, just before the advent of modern whaling: "The moot point is, whether Leviathan can long endure so wide a chase, and so remorseless a havoc; whether he must not at last be exterminated from the waters, and the last whale, like the last man, smoke his last pipe, and then himself evaporate in the final puff."[100] Melville pictured the demise of whaling as a tragic romance. He would have been disappointed to learn that the story ended as a tragic farce and that the last whale smoked its last pipe and then evaporated in the margarine vats of Unilever.

Time Line of Whale Protection

1868   Svend Foyn initiated "modern whaling" in the Varanger fjord of Finnmark in northeastern Norway.

1904   Whaling commenced in South Georgia, opening up the Southern Ocean around the Antarctic Continent.

1909   The first hydrogenation factory was built (based on technology developed in 1903), spurring the demand for whale oil.

1925   The era of "pelagic whaling" began in Antarctica's Ross Sea.

1929   The Norwegian Whaling Act (the 1929 Norwegian Act) was passed by Norway's parliament. It was the first attempt to establish international rules for pelagic whaling.

1931   The Convention for the Regulation of Whaling (the 1931 Geneva Convention) was signed in Geneva on September 24. It was first attempt to regulate the killing of baleen whales. A record number of whales were killed in 1931, mostly in the Antarctic.

1937   The International Agreement for the Regulation of Whaling (the 1937 London Convention) was signed in London on June 8.

1938   The Protocol Amending the International Agreement of June 8, 1937 for the Regulation of Whaling (the 1938 London Protocol) was signed in London on June 24.

1944   The Protocol Amending in Certain Particulars the International Agreement of June 8, 1937 for the Regulation of Whaling (as amended by the Protocol of June 24, 1938) was signed in London on February 7. Because Ireland did not ratify, this protocol never came into force.

1945   The Supplementary Protocol Concerning Whaling was signed in London on October 5. It made the provisions of the 1944 protocol applicable to the 1945–46 season for those countries that signed, thus circumventing Ireland's failure to ratify the agreement.

1945   The Protocol Amending the Agreement of June 8, 1937 for the Regulation of Whaling (as amended by the Protocol of June 24, 1938) was signed in London on November 26. It applied only to the 1946–47 season but was renewed in modified form for the 1947–48 season.

1946   The International Convention for the Regulation of Whaling (the 1946 Washington Convention) was signed in Washington, DC, on December 2. It established the International Whaling Commission (IWC) and set up a permanent regime for regulating whale hunting.

1972   The United Nations Conference on the Human Environment, meeting in Stockholm, called for a ten-year moratorium on whaling.

1973   The Convention on International Trade in Endangered Species (CITES) was signed in Washington, D.C. Most whale species would later receive CITES protection.

1982   The IWC accepted a ten-year moratorium on whaling, though not all whaling nations agreed to adhere to the decision. As of 2009, this moratorium was still in effect.

Conclusion

"CONSERVATION," ALDO Leopold once quipped, "is a bird that flies faster than the shot we aim at it."[1] Mobility was once one of the main advantages that animals had over the humans who hunted them for food and profit, and few animals were faster and more elusive than those that undertook long migrations each year. Snares, traps, arrows, harpoons, stampedes, muskets, and the like certainly took their toll, but the vast majority of species were nimble enough to survive and reproduce, despite some glaring cases of extinction over the centuries. For the past 150 years, however, animals have found themselves staring down the barrel of the modern scientific-industrial revolution. High-powered rifles can strike even the fastest African mammals, just as double-barreled shotguns can devastate entire bird flocks and grenade-tipped harpoons can wipe out entire whale herds. Land usurpation for urban and agricultural development has often proved even more deadly than the weaponry, especially for migratory species, which depend on multiple habitats at various locations during different seasons of the year for their survival. Nothing seems to offer wildlife protection any longer—not the dense forests and vast savannas of Africa, not the wide-open prairies and remote regions of North America, not even the treacherous and frigid polar waters of Antarctica.

During the first half of the twentieth century, animal conservation was all but synonymous with game cropping, multilateral hunting regulations, and nature parks. Diplomatic negotiations were largely in the hands of the major industrial and colonial powers, which responded in ad hoc ways to the problems that arose inside and outside their jurisdiction, be they the overharvesting of ivory, horn, and skins (in Africa); the wanton destructiveness of the meatpacking and millinery industries (in North America); or the exterminationist impulses of the whaling business (in Antarctica). Treaties emerged in piecemeal fashion after one or more governments realized that *national* regulations did not suffice, and the agreements tended to remain *regional* in scope, even if many conservationists dreamed of expanding their geographic domains at some point in the future.

Things began to change after World War II, when the United Nations and other international organizations assumed the task of nature protection worldwide. The new treaties have tended to be global in scope and to codify general rules of conduct for all countries to follow regardless of their geographic circumstances and economic positions. The new treaties also have been far more likely to focus on the protection and restoration of entire ecosystems rather than on the preservation of certain game species and favorite hunting grounds. Three treaties in particular acted as harbingers of a new approach: the 1971 Convention on Wetlands of International Importance Especially as Waterfowl Habitat, signed in Ramsar, Iran (the Ramsar Convention), the first treaty to focus above all on wetland restoration around the globe (even if the term *waterfowl* in its title was a throwback to the old hunting treaties); the 1973 Convention on International Trade in Endangered Species (CITES), signed in Washington, DC, the first treaty designed to rein in the global trade in live animals and animal parts; and the 1979 Convention on the Conservation of Migratory Species of Wild Animals, signed in Bonn (the Bonn Convention), the first treaty devoted solely to the protection of migratory animals and their habitats worldwide.

The old approach to animal protection did not, of course, disappear overnight. The 1968 African Convention largely reiterated the 1933 London Convention, even if it also extended a protective net to endangered native *flora* (an area of conservation that was almost wholly absent from the earlier discussions, since plants were nonmigratory and not a target for hunters). The United States, meanwhile, continued to favor a bilateral approach to bird protection well into the 1970s, signing separate treaties with Japan (1972) and the Soviet Union (1976) that were remarkably similar to the ones signed earlier with Canada and Mexico. Likewise, the IWC has remained the most important agency overseeing the world's whale stocks, even though most whaling enterprises went belly up decades ago for lack of cetaceans to hunt. The new approach to environmental diplomacy did not emerge immediately. The first major conference devoted solely to global environmental issues—the United Nations Conference on the Human Environment, held in Stockholm (the Stockholm Conference)—did not take place until 1972. And the first global environmental agency—the United Nations Environment Programme (UNEP), headquartered in Nairobi, Kenya—was not formally established until a year later. The new diplomacy was born somewhere between the time the last whaling treaty was signed in 1946 and the establishment of UNEP in 1973.

That these newer treaties are, in most respects, superior to the older ones can hardly be denied. British diplomats often casually referred to the 1900 London Convention as the Elephant Treaty, but in reality, it did virtually nothing to protect elephants from ivory predators. The 1933 London Convention also attempted to regulate the African ivory trade, but enforcement in the colonies was far too lax and corruption among colonial officials far too widespread for it to offer much in the way of genuine protection. (Even the much-vaunted British wardens were deeply complicit in the illicit ivory trade, as Ian Parker discovered when he joined the Kenya Game Department in 1956).[2] The novelty of CITES lay in the fact that it placed controls not only on the *export* of tusks and ivory from Africa but also on the *import* of those products into other countries around the globe. In addition, it allowed for a total trade ban if and when a species became endangered, a feature that was missing from the earlier agreements. Rhinos were given comprehensive CITES protection in 1976, elephants in 1989—an accomplishment no earlier treaty was able to achieve.

The Ramsar and Bonn conventions addressed many of the weaknesses that had hampered the 1916 and 1936 bird treaties of North America. The U.S. government's bilateral arrangements were simply too limited in territorial scope to offer comprehensive protection to all of the bird species that migrated across the Western Hemisphere. In practice, the treaties offered more protection to game birds (especially waterfowl) than to other birds, to those that stayed within North America than to those that migrated to Central and South America, and to those that used the Pacific flyway than to those that flew along the Atlantic coast. Both treaties, moreover, lacked effective habitat-protection clauses, and subsequent efforts to rectify this deficiency through the creation of wildlife refuges were never able to keep pace with urban and agricultural development even within North America, let alone elsewhere. The Ramsar Convention dispensed with the bilateral approach altogether in favor of a more global perspective. It focused on identifying and protecting the world's premier wetlands, with the goal of providing all species with adequate feeding and breeding sites. As of 2007, Canada had thirty-seven Ramsar sites covering 150,200 square miles, Mexico had sixty-five sites covering 20,500 square miles, and the United States had twenty-one sites covering 5,020 square miles. Peru had eleven Ramsar sites (26,300 square miles), Bolivia eight (25,100 square miles), Brazil eight (24,700 square miles), Argentina fifteen (13,900 square miles), and Cuba six (4,600 square miles).[3] The Bonn Convention has had a similar impact: it requires countries to set aside sufficient space to accommodate the entire range of an animal's annual movements.

Better protection for whales came in fits and starts, not least because the global commons problem on the high seas had made it difficult even for the United Nations to act effectively. In the early 1960s, the IWC drastically reduced its annual catch quota from 16,000 BWU to 2,700 BWU, a tacit acknowledgment that it had been allowing overharvesting for the previous decade and a half. Then, in 1972, the overwhelming majority of delegates to the Stockholm Conference voted in favor of a complete moratorium on whale hunting for a ten-year period in order to allow the stocks to recover (a recommendation that the IWC temporarily chose to ignore). Shortly thereafter, Greenpeace—an international environmental organization founded in 1971 to halt nuclear testing—began a high-profile campaign against the whaling industry that included film footage of whalers flagrantly violating the terms of the 1946 Washington Convention. At the same time, many nonwhaling nations began to join the IWC (as permitted by Article X of the 1946 treaty) for the sole purpose of putting a complete stop to commercial whaling. They accomplished this mission in 1982, when more than three-fourths of the IWC membership voted to amend the schedule to read: "Catch limits for the killing for commercial purposes of whales from all stocks for the 1986 coastal and the 1985–86 pelagic seasons and thereafter shall be zero."[4] By 1986, CITES protection had been extended to many cetacean species, further ensuring the prospect of a long-term ban on the hunting of the great whales, regardless of what the IWC might decide in the future.

A century ago, governments would have found it politically impossible to impose a total trade ban on elephant ivory and whale oil, so strong was the faith in free enterprise, free trade, and the "free goods" of nature.

But profit-oriented enterprises overharvested these products to such an extent that it is now all but impossible to find diplomats and conservationists willing to sanction even a modicum of trade. African governments with large and stable elephant herds, for instance, have failed in their repeated efforts to revive the ivory trade, even though it makes sense both economically and ecologically to harvest tusks on a sustainable basis. Thus far, CITES members have refused to lift the ivory ban out of fear that it would just reinvigorate the illicit trade in tusks (though in 2002, they did allow South Africa, Botswana, and Namibia to sell sixty tons of ivory from elephants that had died of natural causes). Similarly, Japan, Norway, and the Soviet Union (Russia) officially objected to the IWC's zero quota; under the terms of the 1946 Washington Convention, these objections allow them to ignore the restrictions. The glare of world opinion, however, has so far kept them from resuming full-scale whaling, even though minkes (the smallest of the great whales) are still plentiful in the ocean and several other species have begun to make a rebound. As with ivory, the IWC and CITES members have refused to lift the whaling restriction, largely for fear that doing so would just encourage pirate enterprises. The earlier treaties failed so miserably in their efforts to rein in the elephant-ivory and whale-oil trade that few statespeople are willing to risk a new attempt at game cropping.

To dwell on the weaknesses of the older treaties, however, is to risk losing sight of their many positive features. Africa's national parks and nature reserves are, in many ways, the envy of the world today: they are larger, more conservation-oriented, and more profitable than many of the parks and reserves that were later established in Asia and Latin America, where multilateral agreements were largely lacking. Ivory poaching continues to be a problem, both inside and outside Africa's parks, but here too there is a silver lining: there is no longer much demand for tusks in Europe and the United States, the two regions that once dominated the world's ivory trade. Similarly, North American governments monitor and protect their bird populations better than most other nations of the world, while sustaining the harvesting of millions of game birds each year by recreational hunters. Even the selection of Ramsar sites in North America was facilitated by the fact that many of the prime wetland locations were already under national protection as wildlife and bird refuges. The whaling agreements were far less successful in offering protection to cetaceans, but it is nonetheless worth remembering that they granted protection to the great whales once they became endangered and that all of them survived—no small feat given the rapacious behavior of whaling enterprises (and the continued behavior of Japanese whalers, who invoke the scientific-research clause of the 1946 Washington Convention to justify their annual expeditions).

For all their weaknesses, the treaties discussed in this book helped codify the "rules of the game" that offered many species a sporting chance of survival against the techno-onslaught of the modern era. These treaties marked the beginning point, not the end point, of a long-term diplomatic effort to address the enormous threat that the unrestrained pursuit of profit and the development of ever more efficient killing technologies have posed to the survival of the world's migratory species. Although they did not always provide sustainable long-range solutions, they did at least postpone the day of reckoning long enough for subsequent generations to find more permanent and effective solutions. Sometimes, all that stood between precarious survival and complete extermination was a flawed multinational convention.

Texts of African Treaties

*Convention for the Preservation of Wild
Animals, Birds, and Fish in Africa.
Signed at London, May 19, 1900
(The "1900 London Convention")*

Being desirous of saving from indiscriminate slaughter, and of insuring the preservation throughout their possessions in Africa of the various forms of animal life existing in a wild state which are either useful to man or are harmless, [the parties hereto] have resolved, on the invitation addressed to them by the Government of Her Majesty the Queen of the United Kingdom of Great Britain and Ireland, Empress of India, in accord with the Government of His Majesty the German Emperor, King of Prussia, to assemble with this object a Conference at London . . . [which has] adopted the following provisions:

Article I

The zone within which the provisions of the present Convention shall apply is bounded as follows: On the north by the 20th parallel of north latitude, on the west by the Atlantic Ocean, on the east by the Red Sea and by the Indian Ocean, on the south by a line following the northern boundary of the German possessions in South-Western Africa, from its western extremity to its junction with the River Zambesi, and thence running along the right bank of that river as far as the Indian Ocean.

Article II

The High Contracting Powers declare that the most effective means of preserving the various forms of animal life existing in a wild state within the zone defined in Article I are the following:

1. Prohibition of the hunting and destruction of the animals mentioned in Schedule I attached to the present Convention, and also of any other animals whose protection, whether owing to their usefulness or to their rarity and threatened extermination, may be considered necessary by each Local Government.

2. Prohibition of the hunting and destruction of young animals of the species mentioned in Schedule II attached to the present Convention.

3. Prohibition of the hunting and destruction of females of the species mentioned in Schedule III attached to the present Convention when accompanied by their young. The prohibition, to a certain extent, of the destruction of any females, when they can be recognized as such, with the exception of those of the species mentioned in Schedule V attached to the present Convention.

4. Prohibition of the hunting and destruction, except in limited numbers, of animals of the species mentioned in Schedule IV attached to the present Convention.

5. Establishment, as far as it is possible, of reserves within which it shall be unlawful to hunt, capture, or kill any bird or other wild animal except those which shall be specially exempted from protection by the local authorities. By the term "reserves" are to be understood sufficiently large tracts of land which have all the qualifications necessary as regards food, water, and, if possible, salt, for preserving birds or other wild animals, and for affording them the necessary quiet during the breeding time.

6. Establishment of close seasons with a view to facilitate the rearing of young.

7. Prohibition of the hunting of wild animals by any persons except holders of licences issued by the Local Government, such licences to be revocable in case of any breach of the provisions of the present Convention.

8. Restriction of the use of nets and pitfalls for taking animals.

9. Prohibition of the use of dynamite or other explosives, and of poison, for the purpose of taking fish in rivers, streams, brooks, lakes, ponds, or lagoons.

10. Imposition of export duties on the hides and skins of giraffes, antelopes, zebras, rhinoceroses, and hippopotami, on rhinoceros and antelope horns, and on hippopotamus tusks.

11. Prohibition of hunting or killing young elephants, and, in order to insure the efficacy of this measure, establishment of severe penalties against the hunters, and the confiscation in every case, by the Local Governments, of all elephant tusks weighing less than 5 kilogrammes. The confiscation shall not be enforced when it shall be duly proved that the possession of the tusks weighing less than 5 kilogrammes was anterior to the date of the coming into force of the present Convention. No such proof shall be accepted a year after that date.

12. Application of measures, such as the supervision of sick cattle, &c., for preventing the transmission of contagious diseases from domestic animals to wild animals.

13. Application of measures for effecting the sufficient reduction of the numbers of the animals of the species mentioned in Schedule V attached to the present Convention.

14. Application of measures for insuring the protection of the eggs of ostriches.

15. Destruction of the eggs of crocodiles, of those of poisonous snakes, and of those of pythons.

Article III

The Contracting Parties undertake to promulgate, within a year from the date on which the present Convention comes into force, unless they already exist, provisions applying in their respective possessions within the zone defined in Article I the principles and measures laid down in Article II, and to communicate to one another, as soon as possible after issue, the text of such provisions, and, within eighteen months, information as to the areas which may be established as reserves.

It is, however, understood that the principles laid down in paragraphs 1, 2, 3, 5, and 9 of Article II may be relaxed, either in order to permit the collection of specimens for museums or zoological gardens, or for any other scientific purpose, or in cases where such relaxation is desirable for important administrative reasons, or necessitated by temporary difficulties in the administrative organization of certain territories.

Article IV

The Contracting Parties undertake to apply, as far as possible, each in their respective possessions, measures for encouraging the domestication of zebras, of elephants, of ostriches, &c.

Article V

The Contracting Parties reserve to themselves the right to introduce into the present Convention, by common accord, such modifications or improvements as experience may show to be useful.

Article VI

The Powers having territories or possessions within the zone defined in Article I, who have not signed the present Convention, shall be permitted to accede to it. With this object, the Government of Her Britannic Majesty is charged to communicate the present Convention to them before the exchange of ratifications.

The accession of each Power shall be notified through the diplomatic channel to the Government of Her Britannic Majesty, and by that Government to all the signatory or acceding States.

Such accession shall of itself carry with it acceptance of all the obligations stipulated in the present Convention.

Article VII

The Contracting Parties reserve to themselves the right to introduce, or to propose to the Legislatures of their self-governing Colonies, the necessary measures for carrying out the stipulations of the present Convention in their possessions and Colonies contiguous to the zone defined by Article I.

Article VIII

The present Convention shall be ratified.

The ratifications shall be deposited in London as soon as possible, and shall remain deposited in the archives of the Government of Her Britannic Majesty.

As soon as all the ratifications shall have been produced, a Protocol of deposit shall be drawn up which shall be signed by the Representatives in London of the Powers who shall have ratified.

A certified copy of this Protocol shall be forwarded to each of the Powers interested.

Article IX

The present Convention shall come into force one month after the date of the signature of the Protocol of deposit of the ratifications provided for in Article VIII.

Article X

The present Convention shall remain in force for fifteen years, and in the event of none of the Contracting Parties having notified, twelve months before the expiration of the said period of fifteen years, its intention of terminating its operation, it shall continue to remain in force for a year, and so on from year to year.

In case one of the signatory or acceding Powers shall denounce the Convention, such denunciation shall only affect the Power in question.

In witness whereof the respective Plenipotentiaries have signed the present Convention, and have thereto affixed their seals.

Done at London, in septuplicate, one copy for each Party, the nineteenth day of the month of May, in the year of our Lord one thousand nine hundred.

Hopetoun
Clement Ll. Hill
E. Ray Lankester
G. v. Lindenfels
Dr. von Wissmann
Pedro Jover y Tovar

F. Fuchs
Geoffray
L. G. Binger
Costa
Jayme Batalha-Reis

Annex

Schedule I.

Animals referred to in paragraph 1 of Article II, whose preservation it is desired to ensure:

(Series A).—On account of their usefulness:

1. Vultures.
2. The Secretary-bird.
3. Owls.
4. Rhinoceros-birds or Beef-eaters (*Buphaga*).

(Series B.)—On account of their rarity and threatened extermination:

1. The Giraffe.
2. The Gorilla.
3. The Chimpanzee.
4. The Mountain Zebra.
5. Wild Asses.
6. The White-tailed Gnu (*Connochaetes gnu*).
7. Elands (*Taurotragus*).
8. The little Liberian Hippopotamus.

Schedule II.

Animals referred to in paragraph 2 of Article II, of which it is desired to prohibit the destruction when young:

1. The Elephant.
2. Rhinoceroses.
3. The Hippopotamus.
4. Zebras of the species not referred to in Schedule I.
5. Buffaloes.
6. Antelopes and Gazelles, especially species of the genera *Bubalis, Damaliscus, Connochaetes, Cephalophus, Oreotragus, Oribia, Rhaphiceros, Nesotragus, Madoqua, Cobus, Cervicapra, Pelea, Aepyceros,*

Antidorcas, Gazella, Ammodorcas, Lithocranius, Dorcotragus, Oryx, Addax, Hippotragus, Taurotragus, Strepsiceros, Tragelaphus.
7. Ibex.
8. Chevrotains (*Tragulus*).

Schedule III.

Animals referred to in paragraph 3 of Article II, the killing of the females of which, when accompanied by their young, is prohibited:
1. The Elephant.
2. Rhinoceroses.
3. The Hippopotamus.
4. Zebras of the species not referred to in Schedule I.
5. Buffaloes.
6. Antelopes and Gazelles, especially species of the genera *Buba-lis, Damaliscus, Connochaetes, Cephalophus, Oreotragus, Oribia, Rhaphiceros, Nesotragus, Madoqua, Cobus, Cervicapra, Pelea, Aepyceros, Antidorcas, Gazella, Ammodorcas, Lithocranius, Dorcotragus, Oryx, Addax, Hippotragus, Taurotragus, Strepsiceros, Tragelaphus.*
7. Ibex.
8. Chevrotains (*Tragulus*).

Schedule IV.

Animals referred to in paragraph 4 of Article II, of which only limited numbers may be killed:
1. The Elephant.
2. Rhinoceroses.
3. The Hippopotamus.
4. Zebras of the species not referred to in Schedule I.
5. Buffaloes.
6. Antelopes and Gazelles, especially species of the genera *Buba-lis, Damaliscus, Connochaetes, Cephalophus, Oreotragus, Oribia, Rhaphiceros, Nesotragus, Madoqua, Cobus, Cervicapra, Pelea, Aepyceros, Antidorcas, Gazella, Ammodorcas, Lithocranius, Dorcotragus, Oryx, Addax, Hippotragus, Taurotragus, Strepsiceros, Tragelaphus.*
7. Ibex.
8. Chevrotains (*Tragulus*).
9. The various Pigs.
10. Colobi and all the fur-Monkeys.
11. Aard-Varks (genus *Orycteropus*).

12. Dugongs (genus *Halicore*).
13. Manatees (genus *Manatus*).
14. The small Cats.
15. The Serval.
16. The Cheetah (*Cynalurus*).
17. Jackals.
18. The Aard-wolf (*Proteles*).
19. Small Monkeys.
20. Ostriches.
21. Marabous.
22. Egrets.
23. Bustards.
24. Francolins, Guinea-fowl and other "Game" birds.
25. Large Tortoises.

Schedule V.

Harmful animals referred to in paragraphs 3 and 13 of Article II, of which it is desired to reduce the numbers within sufficient limits:

1. The Lion.
2. The Leopard.
3. Hyaenas.
4. The Hunting Dog (*Lycaon pictus*).
5. The Otter (*Lutra*).
6. Baboons (*Cynocephalus*) and other harmful Monkeys.
7. Large birds of prey, except Vultures, the Secretary-bird and Owls.
8. Crocodiles.
9. Poisonous Snakes.
10. Pythons.

Convention Relative to the Preservation of Fauna and Flora in their Natural State. Signed in London on November 8, 1933 (The "1933 London Convention")

The Governments of the Union of South Africa, Belgium, the United Kingdom of Great Britain and Northern Ireland, Egypt, Spain, France, Italy, Portugal, and the Anglo-Egyptian Sudan:

Considering that the natural fauna and flora of certain parts of the world, and in particular of Africa, are in danger, in present conditions, of extinction or permanent injury;

Desiring to institute a special régime for the preservation of fauna and flora;

Considering that such preservation can best be achieved (i) by the constitution of national parks, strict natural reserves, and other reserves within which the hunting, killing or capturing of fauna, and the collection or destruction of flora shall be limited or prohibited, (ii) by the institution of regulations concerning the hunting, killing and capturing of fauna outside such areas, (iii) by the regulation of the traffic in trophies, and (iv) by the prohibition of certain methods of and weapons for the hunting, killing and capturing of fauna;

Have decided to conclude a Convention for these purposes . . . and have agreed on the following provisions:

Article I

1. Save as regard the territories mentioned in paragraph 3 (i) of the present Article, any Contracting Government shall be at liberty, in accordance with the provisions of Article 13, to assume, in respect of any of its territories (including metropolitan territories, colonies, overseas territories, or territories under suzerainty, protection, or mandate), only those obligations of the present Convention which are set out in Article 9, paragraphs 3, 8 and 9. The term "in part" in the present Convention shall be deemed to refer to those obligations.

2. The expression "territory" or "territories" in relation to any Contracting Government shall, for the purposes of Articles 2–12 of the present Convention, denote the territory or territories of that Government to which the Convention is applicable in full; and, subject to the provisions of the preceding paragraph and of Article 13, the obligations arising under Articles 2–12 shall relate only to such territories.

3. The present Convention shall apply and shall be applicable in full to (i) all the territories (i.e., metropolitan territories, colonies, overseas territories, or territories under suzerainty, protection, or mandate) of any Contracting Government which are situated in the continent of Africa, including Madagascar and Zanzibar; (ii) any other terri-

tory in respect of which a Contracting Government shall have assumed all the obligations of the present Convention in accordance with the provisions of Article 13.

4.	For the purposes of the present Convention the British High Commission Territories in South Africa shall be regarded as a single territory.

5.	The present Convention shall not have any application, either in full or in part, to any metropolitan territory not situated in the continent of Africa, except where and to the extent to which a declaration effecting such application is made under Article 13.

Article 2

For the purposes of the present Convention:

1.	The expression "national park" shall denote an area (a) placed under public control, the boundaries of which shall not be altered or any portion be capable of alienation except by the competent legislative authority, (b) set aside for the propagation, protection and preservation of wild animal life and wild vegetation, and for the preservation of objects of esthetic, geological, prehistoric, historical, archaeological, or other scientific interest for the benefit, advantage, and enjoyment of the general public, (c) in which the hunting, killing or capturing of fauna and the destruction or collection of flora is prohibited except by or under the direction or control of the park authorities.

	In accordance with the above provisions facilities shall, so far as possible, be given to the general public for observing the fauna and flora in national parks.

2.	The term "strict natural reserve" shall denote an area placed under public control, throughout which any form of hunting or fishing, any undertakings connected with forestry, agriculture, or mining, any excavations or prospecting, drilling, levelling of the ground, or construction, any work involving the alteration of the configuration of the soil or the character of the vegetation, any act likely to harm or disturb the fauna or flora, and the introduction of any species of fauna and flora, whether indigenous or imported, wild or domesticated, shall be strictly forbidden; which it shall be forbidden to enter, traverse, or camp in without a special written permit from the competent authorities; and in which scientific investigations may only be undertaken by permission of those authorities.

3. The expression "animal" or "species" shall denote all vertebrates and invertebrates (including non-edible fish, but not including edible fish except in a national park or strict natural reserve), their nests, eggs, egg-shells, skins, and plumage.

Article 3

1. The Contracting Governments will explore forthwith the possibility of establishing in their territories national parks and strict natural reserves as defined in the preceding Article. In all cases where the establishment of such parks or reserves is possible, the necessary work shall be commenced within two years from the date of the entry into force of the present Convention.

2. If in any territory the establishment of a national park or strict natural reserve is found to be impracticable at present, suitable areas shall be selected as early as possible in the development of the territory concerned, and the areas so selected shall be transformed into national parks or strict natural reserves so soon as, in the opinion of the authorities of the territory, circumstances will permit.

Article 4

The Contracting Governments will give consideration in respect of each of their territories to the following administrative arrangements:

1. The control of all white or native settlements in national parks with a view to ensuring that as little disturbance as possible is occasioned to the natural fauna and flora.

2. The establishment round the borders of national parks and strict natural reserves of intermediate zones within which the hunting, killing and capturing of animals may take place under the control of the authorities of the park or reserve; but in which no person who becomes an owner, tenant, or occupier after a date to be determined by the authority of the territory concerned shall have any claim in respect of depredations caused by animals.

3. The choice in respect of all national parks of areas sufficient in extent to cover, so far as possible, the migrations of the fauna preserved therein.

Article 5

1. The Contracting Governments shall notify the Government of the United Kingdom of Great Britain and Northern Ireland of the establishment of any national parks or strict natural reserves (defining the area of the parks or reserves), and of the legislation, including the methods of administration and control, adopted in connexion therewith.

2. They shall similarly notify any information relevant to the purposes of the present Convention and communicated to them by any national museums or by any societies, national or international, established within their jurisdiction and interested in those purposes.

3. The Government of the United Kingdom will communicate the information so received to the other Governments which have signed or acceded to the present Convention whether in full or in part.

Article 6

In all cases in which it is proposed to establish in any territory of a Contracting Government a national park or strict natural reserve contiguous to a park or reserve situated in another territory (whether of that Government or of another Contracting Government), or to the boundary of such territory, there shall be prior consultation between the competent authorities of the territories concerned. Similarly, there shall be co-operation between those authorities subsequent to the establishment of the park or reserve, or where such a park or reserve is already established.

Article 7

Irrespective of any action which may be taken under Article 3 of the present Convention, the Contracting Governments shall, as measures preliminary and supplementary to the establishment of national parks or strict natural reserves:

1. Set aside in each of their territories suitable areas (to be known as reserves) within which the hunting, killing, or capturing of any part of the natural fauna (exclusive of fish) shall be prohibited save (a) by the permission, given for scientific or administrative purposes in exceptional cases by the authorities of the territory or by the central authorities under whom the reserves are placed or (b) for the

protection of life and property. Licences granted under Article 8, paragraphs 1 and 3, shall not extend to reserves.

2. Extend in these areas, so far as may be practicable, a similar degree of protection to the natural flora.

3. Consider the possibility of establishing in each of their territories special reserves for the preservation of species of fauna and flora which it is desired to preserve, but which are not otherwise adequately protected, with special reference to the species mentioned in the Annex to the present Convention.

4. Furnish information regarding the reserves established in accordance with the preceding paragraphs to the Government of the United Kingdom, which will communicate such information to all the Governments mentioned in Article 5, paragraph 2.

5. Take, so far as in their power lies, all necessary measures to ensure in each of their territories a sufficient degree of forest country and the preservation of the best native indigenous forest species, and, without prejudice to the provisions of Article 2, paragraph 2, give consideration to the desirability of preventing the introduction of exotic trees or plants into national parks or reserves.

6. Establish as close a degree of co-operation as possible between the competent authorities of their respective territories with the object of facilitating the solution of forestry problems in those territories.

7. Take the necessary measures to control and regulate so far as possible the practice of firing the bush on the borders of forests.

8. Encourage the domestication of wild animals susceptible of economic utilisation.

Article 8

1. The protection of the species mentioned in the Annex to the present Convention is declared to be of special urgency and importance. Animals belonging to the species mentioned in Class A shall, in each of the territories of the Contracting Governments, be protected as completely as possible, and the hunting, killing, or capturing of them shall only take place by special permission of the highest authority in the territory, which shall be given only under

special circumstances, solely in order to further important scientific purposes, or when essential for the administration of the territory. Animals belonging to the species mentioned in Class B, whilst not requiring such rigorous protection as those mentioned in Class A, shall not be hunted, killed, or captured, even by natives, except under special licence granted by the competent authorities. For this purpose a special licence shall denote a licence other than an ordinary game licence, granted at the discretion of the competent authority, and giving permission to hunt, kill, or capture one or more specimens of a specified animal or animals. Every such licence shall be limited as regards the period and the area within which hunting, killing, or capturing may take place.

2. No hunting or other rights already possessed by native chiefs or tribes or any other persons or bodies, by treaty, concession, or specific agreement, or by administrative permission in those areas in which such rights have already been definitely recognised by the authorities of the territory are to be considered as being in any way prejudiced by the provisions of the preceding paragraph.

3. In each of the territories of the Contracting Governments the competent authorities shall consider whether it is necessary to apply the provisions of paragraph 1 of the present Article to any species not mentioned in the Annex, in order to preserve the indigenous fauna or flora in each area, and, if they deem it necessary, shall apply those provisions to any such species to the extent which they consider desirable. They shall similarly consider whether it is necessary in the territory concerned to accord to any of the species mentioned in Class B of the Annex the special protection accorded to the species mentioned in Class A.

4. The competent authorities shall also give consideration to the question of protecting species of animals or plants which by general admission are useful to man or of special scientific interest.

5. Nothing in the present Article shall (i) prejudice any right which may exist under the local law of any territory to kill animals without a licence in defence of life or property, or (ii) affect the right of the authorities of the territory to permit the hunting, killing, or capturing of any species (a) in time of famine, (b) for the protection of human life, public health, or domestic stock, (c) for any requirement relating to public order.

6.   Each Contracting Government shall furnish to the Government of the United Kingdom information on the subject of the measures adopted in each of its territories in regard to the grant of licences, and in regard to the animals, the destruction or capture of which is, in accordance with paragraph 3 of this Article, not permitted except under licence. The Government of the United Kingdom will communicate any such information to all the Governments mentioned in Article 5, paragraph 2.

Article 9

1.   Each Contracting Government shall take the necessary measures to control and regulate in each of its territories the internal, and the import and export, traffic in, and the manufacture of articles from, trophies as defined in paragraph 8 of the present Article, with a view to preventing the import or export of, or any dealing in, trophies other than such as have been originally killed, captured or collected in accordance with the laws and regulations of the territory concerned.

2.   The export of trophies to any destination whatsoever shall be prohibited unless the exporter has been granted a certificate permitting export and issued by a competent authority. Such certificate shall only be issued where the trophies have been lawfully imported or lawfully obtained. In the event of an attempted export without any certificate having been granted, the authorities of the territory where this attempt takes place shall apply such penalties as they may think necessary.

3.   The import of trophies which have been exported from any territory to which the present Convention is applicable in full, whether a territory of another Contracting Government or not, shall be prohibited except on production of a certificate of lawful export, failing which the trophy shall be confiscated, but without prejudice to the application of the penalties mentioned in the preceding paragraph.

4.   The import and export of trophies, except at places where there is a Customs station, shall be prohibited.

5.   (a) Every trophy consisting of ivory and rhinoceros horn exported in accordance with the provisions of the present Article shall be

identified by marks which, together with the weight of the trophy, shall be recorded in the certificate of lawful export.

(b) Every other trophy shall, if possible, be similarly marked and recorded but shall in any event be described in the certificate so as to identify it with as much certainty as possible.

(c) The Contracting Governments shall take such measures as may be possible by the preparation and circulation of appropriate illustrations or otherwise to instruct their Customs officers in the methods of identifying the species mentioned in the Annex to the present Convention and the trophies derived therefrom.

6. The measures contemplated in paragraph 1 of the present Article shall include provisions that found ivory, rhinoceros horn and all trophies of animals found dead, or accidentally killed, or killed in defence of any person, shall, in principle, be the property of the Government of the territory concerned, and shall be disposed of according to regulations introduced by that Government, due regard being had to the native rights and customs reserved in the succeeding paragraph.

7. No rights of the kind specified in paragraph 2 of Article 8 are to be considered as being prejudiced by the provisions of the preceding paragraphs.

8. For the purposes of the present Article the expression "trophy" shall denote any animal, dead or alive, mentioned in the Annex to the Convention, or anything part of or produced from any such animal when dead, or the eggs, egg-shells, nest or plumage of any bird so mentioned. The expression "trophy" shall not, however, include any trophy or part of a trophy which by a process of *bona fide* manufacture, as contemplated in paragraph 1 of the present Article, has lost its original identity.

9. Each Contracting Government shall furnish to the Government of the United Kingdom information as to the measures taken in order to carry out the obligations of the present Article or any part of them. The Government of the United Kingdom will communicate any information so received to all the Governments mentioned in Article 5, paragraph 2.

Article 10

1. The use of motor vehicles or aircraft (including aircraft lighter than air) shall be prohibited in the territories of the Contracting Governments, both (i) for the purpose of hunting, killing, or capturing animals, and (ii) in such manner as to drive, stampede, or disturb them for any purpose whatsoever, including that of filming or photographing; provided, however, that nothing in the present paragraph shall affect the right of occupiers in respect of land occupied by them, or of Governments in respect of land utilised for public purposes, to use motor vehicles or aircraft for the purpose of driving away, capturing, or destroying animals found on such land in all cases where such ejection, capture, or destruction is not prohibited by any other provision of the present Convention.

2. The Contracting Governments shall prohibit in their territories the surrounding of animals by fires for hunting purposes. Wherever possible, the under-mentioned methods of capturing or destroying animals shall also be generally prohibited:

   (a) The use of poison, or explosives for killing fish;

   (b) The use of dazzling lights, flares, poison, or poisoned weapons for hunting animals;

   (c) The use of nets, pits or enclosures, gins, traps or snares, or of set guns and missiles containing explosives for hunting animals.

Article 11

It is understood that upon signature, ratification, or accession any Contracting Government may make such express reservations in regard to Articles 3–10 of the present Convention as may be considered essential.

Article 12

1. Each Contracting Government shall furnish to the Government of the United Kingdom information as to the measures taken for the purpose of carrying out the provisions of the preceding Articles. The Government of the United Kingdom will communicate all the information so furnished to the Governments mentioned in Article 5, paragraph 2.

2. The Contracting Governments shall, wherever necessary, co-operate between themselves for the purpose of carrying out the provisions of the preceding Articles and to prevent the extinction of fauna and flora.

3. All the Governments which sign or accede to the present Convention shall be deemed to be Parties to the Protocol bearing this day's date, drawn up to facilitate the co-operation mentioned in the preceding paragraph.

Article 13

1. Any Contracting Government may, at the time of signature, ratification, or accession, or thereafter, make a declaration assuming in respect of any one or more of its territories (including metropolitan territories, colonies, overseas territories, or territories under suzerainty, protection, or mandate) other than those mentioned in paragraph 3 (i) of Article 1, either all the obligations of the present Convention, or only those contained in Article 9, paragraphs 3, 8 and 9. If such declaration is made subsequent to ratification or accession it shall be effected by means of a notification in writing addressed to the Government of the United Kingdom, and shall take effect on the entry into force of the Convention or, if the Convention is already in force, three months after the date of the receipt of the notification by the Government of the United Kingdom.

2. It is understood that any Contracting Government may, by a single declaration made under the preceding paragraph, assume, in respect of some of its territories mentioned in that paragraph, all the obligations of the present Convention, and in respect of other such territories only the obligations contained in Article 9, paragraphs 3, 8 and 9.

3. Any Contracting Government which has made a declaration under the preceding paragraph, assuming, in respect of any territory, only the obligations contained in Article 9, paragraphs 3, 8 and 9, may, at any subsequent time, by a notification in writing addressed to the Government of the United Kingdom, declare that such previous declaration shall henceforth be deemed to relate to all the obligations of the Convention in respect of the territory concerned; and such subsequent declaration shall take effect on the entry into force of the Convention or, if the Convention is already in force, three months after the date of the receipt of the notification by the Government of the United Kingdom.

4. Any Contracting Government may at any time, by a notification in writing addressed to the Government of the United Kingdom, determine the application of the Convention to any territory or territories which have been the subject of a declaration under paragraphs 1 and 3 of the present Article, and the Convention shall thereupon cease to apply to the territory or territories mentioned in the notification one year after the date of its receipt by the Government of the United Kingdom, provided that such notification shall in no case take effect until the expiry of the period of five years mentioned in Article 19, paragraph 1.

5. It is understood that if, as the result of a notification made under the preceding paragraph, there would remain no territories of the Contracting Government concerned to which the Convention would be applicable either in full or in part, such Government shall, instead of making the notification, proceed by way of denunciation under Article 19.

6. It is further understood that no notification made under paragraph 4 of the present Article, or otherwise, may purport to apply only the provisions of Article 9, paragraphs 3, 8 and 9, to any territory to which, at the time of the notification, the Convention applies in full.

7. The Government of the United Kingdom will inform all the Governments mentioned in Article 5, paragraph 2, of any notifications received under the preceding paragraphs of the present Article, of the date of their receipt and of their terms.

Article 14

It is understood that no Government will sign, ratify, or accede to the present Convention unless it either has territories covered by Article 1, paragraph 3 (i), or makes or has made a declaration under Article 13 assuming in respect of one or more territories the obligations of the Convention either in full or in part.

Article 15

The present Convention, of which the French and English texts shall both be equally authentic, shall bear this day's date and shall be open for signature until the 31st March, 1934.

Article 16

The present Convention shall be subject to ratification. The instruments of ratification shall be deposited with the Government of the United Kingdom, which will notify their receipt and the date thereof, and their terms and terms of any accompanying declarations or reservations to all Governments mentioned in Article 5, paragraph 2.

Article 17

At any time after the 31st March, 1934, the present Convention shall be open to accession by any Government of a metropolitan territory, by which it has not been signed, whether it has territories covered by Article 1, paragraph 3 (i), or not. Accessions shall be notified to the Government of the United Kingdom, which will inform all the Governments mentioned in Article 5, paragraph 2, of all notifications received, their terms and the terms of any accompanying declarations or reservations, and the date of their receipt.

Article 18

1. After the deposit or notification of not less than four ratifications or accessions on the part of Contracting Governments having territories covered by Article 1, paragraph 3 (i), the present Convention shall come into force three months after the deposit or notification of the last of such ratifications or accessions, as between the Governments concerned. The Government of the United Kingdom will notify all the Governments mentioned in Article 5, paragraph 2, of the date of the coming into force of the Convention.

2. Any ratifications or accessions received after the date of the entry into force of the Convention shall take effect three months after the date of their receipt by the Government of the United Kingdom.

Article 19

1. Any Contracting Government may at any time denounce the present Convention by a notification in writing addressed to the Government of the United Kingdom. Such denunciation shall take effect, as regards the Government making it, and in respect of all

the territories of that Government to which the Convention shall then apply, either in full or in part, one year after the receipt of the notification by the Government of the United Kingdom; provided, however, that no denunciation shall take effect until the expiry of five years from the date of the entry into force of the Convention.

2. If, as the result of simultaneous or successive denunciations, the number of Contracting Governments bound, in respect of one or more of their territories, by all the obligations of the present Convention is reduced to less than four, the Convention shall cease to be in force as from the date on which the last of such denunciations shall take effect in accordance with the provisions of the preceding paragraph.

3. The Government of the United Kingdom will notify all the other Governments mentioned in Article 5, paragraph 2, of any denunciations so received and the date on which they take effect. The Government of the United Kingdom will also, if occasion arises, similarly notify the date on which the Convention ceases to be in force under the provisions of the preceding paragraph.

In witness whereof the above-named Plenipotentiaries have signed the present Convention.

Done in London, this eighth day of November, 1933, in a single copy, which shall remain deposited in the archives of the Government of the United Kingdom of Great Britain and Northern Ireland, which will transmit certified true copies thereof to all the Governments attending the Conference at which the present Convention has been drawn up, whether as participators or observers, as well as to any other Government to which the Government of the United Kingdom may deem it desirable to communicate a copy.

> *Union of South Africa*
> *Belgium*
> *Great Britain and Northern*
> *Ireland*
> *Egypt*
> *Spain*
> *France*
> *Italy*
> *Portugal*
> *Anglo-Egyptian Sudan*

Annex

Class A

1. Animalia

Gorilla
Madagascar Lemurs
Aard Wolf
Fossa
Giant Sable Antelope
Nyala or Inyala
Mountain Nyala or Buxton's Bushbuck
Okapi
Barbary Stag
Pigmy Hippopotamus
Mountain Zebra
Wild Ass
White Rhinoceros
Northern Hartebeest or Bubal
Abyssinian Ibex or Wali
African Elephant (with tusks under 5 kilograms)
Water Chevrotain
Whale-headed Stork or Shoe-bill
Bald-headed Ibis or Waldrapp
White-breasted Guinea Fowl

2. Vegetabilia

Welwitschia

Class B

1. Animalia

Chimpanzee
Colobus Monkey
Giant Eland or Lord Derby's Eland
Giraffe
White-tailed Gnu
Yellow-backed Duiker
Jentink's Duiker

Beira
Dibatag or Clarke's Gazelle
Bontebok
Black Rhinoceros
African Elephant (with tusks over 5 kilograms)
Pangolin
Marabou
Abyssinian Ground Hornbill
Ground Hornbill
Wild Ostrich
Secretary Bird
Little Egret
African Great White Egret
African Yellow-billed Egret
Buff-backed Egret

Texts of Bird Treaties

Convention for the Protection of Migratory Birds Signed between the United States and Great Britain. Signed in Washington, DC on August 16, 1916 (The "1916 Convention")

Whereas, Many species of birds in the course of their annual migrations traverse certain parts of the United States and the Dominion of Canada; and

Whereas, Many of these species are of great value as a source of food or in destroying insects which are injurious to forests and forage plants on the public domain, as well as to agricultural crops, in both the United States and Canada, but are nevertheless in danger of extermination through lack of adequate protection during the nesting season or while on their way to and from their breeding grounds;

The United States of America and His Majesty the King of the United Kingdom of Great Britain and Ireland and of the British Dominions beyond the Seas, Emperor of India, being desirous of saving from indiscriminate slaughter and of insuring the preservation of such migratory birds as are either useful to man or are harmless, have resolved to adopt some uniform system of protection which shall effectively accomplish such objects and to the end of concluding a convention for this purpose have appointed as their respective Plenipotentiaries:

The President of the United States of America, Robert Lansing, Secretary of State of the United States; and

His Britannic Majesty, the Right Honorable Sir Cecil Arthur Spring Rice, G. C. V. O., K. C. M. G., etc., His Majesty's Ambassador Extraordinary and Plenipotentiary at Washington;

Who, after having communicated to each other their respective full powers which were found to be in due and proper form, have agreed to and adopted the following articles:

Article I

The High Contracting Powers declare that the migratory birds included in the terms of this Convention shall be as follows:

1. Migratory Game Birds:

    (a) Anatidae or waterfowl, including brant, wild ducks, geese, and swans.

    (b) Gruidae or cranes, including little brown, sandhill, and whooping cranes.

    (c) Rallidae or rails, including coots, gallinules and sora and other rails.

    (d) Limicolae or shorebirds, including avocets, curlew, dowitchers, godwits, knots, oyster catchers, phalaropes, plovers, sandpipers, snipe, stilts, surf birds, turnstones, willet, woodcock and yellowlegs.

    (e) Columbidae or pigeons, including doves and wild pigeons.

2. Migratory Insectivorous Birds:

    Bobolinks, catbirds, chicadees, cuckoos, flickers, flycatchers, grosbeaks, humming birds, kinglets, martins, meadowlarks, nighthawks or bull bats, nut-hatches, orioles, robins, shrikes, swallows, swifts, tanagers, titmice, thrushes, vireos, warblers, wax-wings, whippoorwills, woodpeckers and wrens, and all other perching birds which feed entirely or chiefly on insects.

3. Other Migratory Nongame Birds:

    Auks, auklets, bitterns, fulmars, gannets, grebes, guillemots, gulls, herons, jaegers, loons, murres, petrels, puffins, shearwaters, and terns.

Article II

The High Contracting Powers agree that, as an effective means of preserving migratory birds there shall be established the following close seasons during which no hunting shall be done except for scientific or propagating purposes under permits issued by proper authorities.

1. The close season on migratory game birds shall be between March 10 and September 1, except that the close season on the Limicolae or shorebirds in the Maritime Provinces of Canada and in those States of the United States bordering on the Atlantic Ocean which are situated wholly or in part north of Chesapeake Bay shall be between February 1 and August 15, and that Indians may take at any time scoters for food but not for sale. The season for hunting shall be further restricted to such period not exceeding three and one-half months as the High Contracting Powers may severally deem appropriate and define by law or regulation.

2. The close season on migratory insectivorous birds shall continue throughout the year.

3. The close season on other migratory nongame birds shall continue throughout the year, except that Eskimos and Indians may take at any season auks, auklets, guillemots, murres and puffins, and their eggs for food and their skins for clothing, but the birds and eggs so taken shall not be sold or offered for sale.

Article III

The High Contracting Powers agree that during the period of ten years next following the going into effect of this Convention there shall be a continuous close season on the following migratory game birds, to wit:

Band-tailed pigeons, little brown, sandhill and whooping cranes, swans, curlew and all shorebirds (except the black-breasted and golden plover, Wilson or jack snipe, woodcock, and the greater and lesser yellowlegs); provided that during such ten years the close seasons on cranes, swans and curlew in the Province of British Columbia shall be made by the proper authorities of that Province within the general dates and limitations elsewhere prescribed in this Convention for the respective groups to which these birds belong.

Article IV

The High Contracting Powers agree that special protection shall be given the wood duck and the eider duck either (1) by a close season extending over a period of at least five years, or (2) by the establishment of refuges, or (3) by such other regulations as may be deemed appropriate.

Article V

The taking of nests or eggs of migratory game or insectivorous or non-game birds shall be prohibited, except for scientific or propagating purposes under such laws or regulations as the High Contracting Powers may severally deem appropriate.

Article VI

The High Contracting Powers agree that the shipment or export of migratory birds or their eggs from any State or Province, during the continuance of the close season in such State or Province, shall be prohibited except for scientific or propagating purposes, and the international traffic in any birds or eggs at such time captured, killed, taken, or shipped at any time contrary to the laws of the State or Province in which the same were captured, killed, taken, or shipped shall be likewise prohibited. Every package containing migratory birds or any parts thereof or any eggs of migratory birds transported, or offered for transportation from the United States into the Dominion of Canada into the United States, shall have the name and address of the shipper and an accurate statement of the contents clearly marked on the outside of such package.

Article VII

Permits to kill any of the above-named birds which, under extraordinary conditions, may become seriously injurious to the agricultural or other interests in any particular community, may be issued by the proper authorities of the High Contracting Powers under suitable regulations prescribed therefor by them respectively, but such permits shall lapse, or may be cancelled, at any time when, in the opinion of said authorities, the particular exigency has passed, and no birds killed under this article shall be shipped, sold or offered for sale.

Article VIII

The High Contracting Powers agree themselves to take, or propose to their respecting appropriate law-making bodies, the necessary measures for insuring the execution of the present Convention.

Article IX

The present Convention shall be ratified by the President of the United States of America, by and with the advice and consent of the Senate thereof, and by His Britannic Majesty. The ratifications shall be exchanged at Washington as soon as possible and the Convention shall take effect on the date of the exchange of the ratifications. It shall remain in force for fifteen years and in the event of neither of the High Contracting Powers having given notification, twelve months before the expiration of said period of fifteen years, of its intention of terminating its operation, the Convention shall continue to remain in force for one year and so on from year to year.

In faith whereof, the respective Plenipotentiaries have signed the present Convention in duplicate and have hereunto affixed their seals.

Done at Washington this sixteenth day of August, one thousand nine hundred and sixteen.

Robert Lansing.
Cecil Spring Rice.

Convention between the United States of America and the United States of Mexico for the Protection of Migratory Birds and Game Mammals. Signed in Mexico City on February 7, 1936 (The "1936 Convention")

Whereas, some of the birds denominated migratory, in their movements cross the United States of America and the United Mexican States, in which countries they live temporarily;

Whereas, it is right and proper to protect the said migratory birds, whatever may be their origin, in the United States of America and the United Mexican States, in order that the species may not be exterminated;

Whereas, for this purpose it is necessary to employ adequate measures which will permit a rational utilization of migratory birds for the purposes of sport as well as for food, commerce and industry;

The Governments of the two countries have agreed to conclude a Convention which will satisfy the above-mentioned need and to that end have appointed as their respective Plenipotentiaries:

The Honorable Josephus Daniels, representing the President of the United States of America, Franklin D. Roosevelt, and

The Honorable Eduardo Hay, representing the President of the United Mexican States, General Lázaro Cárdenas,

Who, having exhibited to each other and found satisfactory their respective full powers, conclude the following Convention:

Article I

In order that the species may not be exterminated, the high contracting parties declare that it is right and proper to protect birds denominated as migratory, whatever may be their origin, which in their movements live temporarily in the United States of America and the United Mexican States, by means of adequate methods which will permit, in so far as the respective high contracting parties may see fit, the utilization of said birds rationally for purposes of sport, food, commerce and industry.

Article II

The high contracting parties agree to establish laws, regulations and provisions to satisfy the need set forth in the preceding Article, including:

(A) The establishment of close seasons, which will prohibit in certain periods of the year the taking of migratory birds, their nests or eggs, as well as their transportation or sale, alive or dead, their products or parts, except when proceeding, with appropriate authorization, from private game farms or when used for scientific purposes, for propagation or for museums.

(B) The establishment of refuge zones in which the taking of such birds will be prohibited.

(C) The limitation of their hunting to four months in each year, as a maximum, under permits issued by the respective authorities in each case.

(D) The establishment of a close season for wild ducks from the tenth of March to the first of September.

(E) The prohibition of the killing of migratory insectivorous birds, except when they become injurious to agriculture and constitute plagues, as well as when they come from reserves or game farms: provided, however, that such birds may be captured alive and used in conformity with the laws of each contracting country.

(F) The prohibition of hunting from aircraft.

Article III

The high contracting parties respectively agree, in addition, not to permit the transportation over the American-Mexican border of migratory birds, dead or alive, their parts or products, without a permit of authorization provided for that purpose by the Government of each country, with the understanding that in the case that the said birds, their parts or products are transported from one country to the other without the stipulated authorization, they will be considered as contraband and treated accordingly.

Article IV

The high contracting parties declare that for the purposes of the present Convention the following birds shall be considered migratory:

Migratory Game Birds.

Familia Anatidae.	Familia Scolopacidae.
Familia Gruidae.	Familia Recurvirostridae.
Familia Rallidae.	Familia Phalaropodidae.
Familia Charadriidae.	Familia Columbidae.

Migratory Non-game Birds.

Familia Cuculidae.	Familia Mimidae.
Familia Caprimulgidae.	Familia Sylviidae.
Familia Micropodidae.	Familia Motacillidae.
Familia Trochilidae.	Familia Bombycillidae.
Familia Picidae.	Familia Ptilogonatidae.

Familia Tyrannidae. Familia Laniidae.
Familia Alaudidae. Familia Vireonidae.
Familia Hirundinidae. Familia Compsothlypidae.
Familia Paridae. Familia Icteridae.
Familia Certhiidae. Familia Thraupidae.
Familia Troglodytidae. Familia Fringillidae.
Familia Turdidae.

Others which the Presidents of the United States of America and the United Mexican States may determine by common agreement.

Article V

The high contracting parties agree to apply the stipulations set forth in Article III with respect to the game mammals which live in their respective countries.

Article VI

This Convention shall be ratified by the high contracting parties in accordance with their constitutional methods and shall remain in force for fifteen years and shall be understood to be extended from year to year if the high contracting parties have not indicated twelve months in advance their intention to terminate it.

The respective plenipotentiaries sign the present Convention in duplicate in English and Spanish, affixing thereto their respective seals, in the City of Mexico, the seventh day of February of 1936.

Josephus Daniels.
Eduardo Hay.

Texts of Whaling Treaties

Convention for the Regulation of Whaling.
Signed in Geneva on September 24, 1931
(The "1931 Geneva Convention")

Article 1

The High Contracting Parties agree to take, within the limits of their respective jurisdictions, appropriate measures to ensure the application of the provisions of the present Convention and the punishment of infractions of the said provisions.

Article 2

The present Convention applies only to baleens or whalebone whales.

Article 3

The present Convention does not apply to aborigines dwelling on the coasts of the territories of the High Contracting Parties provided that:

(1) They only use canoes, pirogues or other exclusively native craft propelled by oars or sails;

(2) They do not carry firearms;

(3) They are not in the employment of persons other than aborigines;

(4) They are not under contract to deliver the products of their whaling to any third person.

Article 4

The taking or killing of right whales, which shall be deemed to include North-Cape whales, Greenland whales, southern right whales, Pacific right whales and southern pigmy right whales, is prohibited.

Article 5

The taking or killing of calves or suckling whales, immature whales, and female whales which are accompanied by calves (or suckling whales) is prohibited.

Article 6

The fullest possible use shall be made of the carcases of whales taken. In particular:

(1) There shall be extracted by boiling or otherwise the oil from all blubber and from the head and the tongue and, in addition, from the tail as far forward as the outer opening of the lower intestine.

The provisions of this sub-paragraph shall apply only to such carcases or parts of carcases as are not intended to be used for human food.

(2) Every factory, whether on shore or afloat, used for treating the carcases of whales shall be equipped with adequate apparatus for the extraction of oil from the blubber, flesh and bones.

(3) In the case of whales brought on shore, adequate arrangements shall be made for utilising the residues after the oil has been extracted.

Article 7

Gunners and crews of whaling vessels shall be engaged on terms such that their remuneration shall depend to a considerable extent upon such factors as the size, species, value and yield of oil of whales taken, and not merely upon the number of whales taken, in so far as payment is made dependent on results.

Article 8

No vessel of any of the High Contracting Parties shall engage in taking or treating whales unless a licence authorising such vessel to engage therein shall have been granted in respect of such vessel by the High Contracting Party, whose flag she flies, or unless her owner or charterer has notified the Government of the said High Contracting Party of his intention to employ her in whaling and has received a certificate of notification from the said Government.

Nothing in this Article shall prejudice the right of any High Contracting Party to require that, in addition, a licence shall be required from his own authorities by every vessel desirous of using his territory or territorial waters for the purposes of taking, landing or treating whales, and such licence may be refused or may be made subject to such conditions as may be deemed by such High Contracting Party to be necessary or desirable, whatever the nationality of the vessel may be.

Article 9

The geographical limits within which the Articles of this Convention are to be applied shall include all the waters of the world, including both the high seas and territorial and national waters.

Article 10

1. The High Contracting Parties shall obtain, with regard to the vessels flying their flags and engaged in the taking of whales, the most complete biological information practicable with regard to each whale taken, and in any case on the following points:

    (a) Date of taking;

    (b) Place of taking;

    (c) Species;

    (d) Sex;

    (e) Length; measured, when taken out of water; estimated, if cut up in water;

    (f) When foetus is present, length and sex if ascertainable;

    (g) When practicable, information as to stomach contents.

2. The length referred to in sub-paragraphs (e) and (f) of this Article shall be the length of a straight line taken from the tip of the snout to the notch between the flukes of the tail.

Article 11

Each High Contracting Party shall obtain from all factories, on land or afloat, under his jurisdiction, returns of the number of whales of each species treated at each factory and of the amounts of oil of each grade and the quantities of meal, guano and other products derived from them.

Article 12

Each of the High Contracting Parties shall communicate statistical information regarding all whaling operations under their jurisdiction to the International Bureau for Whaling Statistics at Oslo. The information given shall comprise at least the particulars mentioned in Article 10 and: (1) the name and tonnage of each floating factory; (2) the number and aggregate tonnage of the whale catchers; (3) a list of the land stations which were in operation during the period concerned. Such information shall be given at convenient intervals not longer than one year.

Article 13

The obligation of a High Contracting Party to take measures to ensure the observance of the conditions of the present Convention in his own territories and territorial waters, and by his vessels, shall not apply to those of his territories to which the Convention does not apply, and the territorial waters adjacent thereto, or to vessels registered in such territories.

Article 14

The present Convention, the French and English texts of which shall both be authoritative, shall remain open until the thirty-first of March 1932 for signature on behalf of any Member of the League of Nations or of any non-member State.

Article 15

The present Convention shall be ratified. The instruments of ratification shall be deposited with the Secretary-General of the League of Nations, who shall notify their receipt to all Members of the League of Nations and non-member States indicating the dates of their deposit.

Article 16

As from the first of April 1932, any Member of the League of Nations and any non-member State, on whose behalf the Convention has not been signed before that date, may accede thereto.

The instruments of accession shall be deposited with the Secretary-General of the League of Nations, who shall notify all the Members of the League of Nations and non-member States of their deposit and the date thereof.

Article 17

The present Convention shall enter into force on the ninetieth day following the receipt by the Secretary-General of the League of Nations of ratifications or accessions on behalf of not less than eight Members of the League or non-member States, including the Kingdom of Norway and the United Kingdom of Great Britain and Northern Ireland.

As regards any Member of the League or non-member State on whose behalf an instrument of ratification or accession is subsequently deposited, the Convention shall enter into force on the ninetieth day after the date of the deposit of such instrument.

Article 18

If after the coming into force of the present Convention the Council of the League of Nations, at the request of any two Members of the League or non-member States with regard to which the Convention is then in force, shall convene a Conference for the revision of the Convention, the High Contracting Parties agree to be represented at any Conference so convened.

Article 19

1. The present Convention may be denounced after the expiration of three years from the date of its coming into force.

2. Denunciation shall be effected by a written notification addressed to the Secretary-General of the League of Nations, who shall inform

all the Members of the League and the non-member States of each notification received and of the date of its receipt.

3. Each denunciation shall take effect six months after the receipt of its notification.

Article 20

1. Any High Contracting Party may, at the time of signature, ratification or accession, declare that, in accepting the present Convention, he does not assume any obligations in respect of all or any of his colonies, protectorates, overseas territories or territories under suzerainty or mandate; and the present Convention shall not apply to any territories named in such declaration.

2. Any High Contracting Party may give notice to the Secretary-General of the League of Nations at any time subsequently that he desires that the Convention shall apply to all or any of his territories which have been made the subject of a declaration under the preceding paragraph, and the Convention shall apply to all the territories named in such notice ninety days after its receipt by the Secretary-General of the League of Nations.

3. Any High Contracting Party may, at any time after the expiration of the period of three years mentioned in Article 19, declare that he desires that the present Convention shall cease to apply to all or any of his colonies, protectorates, overseas territories or territories under suzerainty or mandate and the Convention shall cease to apply to the territories named in such declaration six months after its receipt by the Secretary-General of the League of Nations.

4. The Secretary-General of the League of Nations shall communicate to all the Members of the League of Nations and the non-member States all declarations and notices received in virtue of this Article and the dates of their receipt.

Article 21

The present Convention shall be registered by the Secretary-General of the League of Nations as soon as it has entered into force.

In faith whereof the above-mentioned Plenipotentiaries have signed the present Convention.

Done at Geneva, on the twenty-fourth day of September one thousand nine hundred and thirty-one, in a single copy which shall be kept in the archives of the Secretariat of the League of Nations and of which certified true copies shall be delivered to all the Members of the League of Nations and to the non-member States.

Albania	*Finland*
Germany	*France*
United States of America	*Greece*
Belgium	*Italy*
Great Britain and Northern Ireland	*Mexico*
Canada	*Norway*
Commonwealth of Australia	*The Netherlands*
New Zealand	*Poland*
Union of South Africa	*Roumania*
India	*Switzerland*
Colombia	*Czechoslovakia*
Denmark	*Turkey*
Spain	*Yugoslavia*

International Agreement for the Regulation of Whaling. Signed in London on June 8, 1937 (The "1937 London Convention")

The Governments of the Union of South Africa, the United States of America, the Argentine Republic, the Commonwealth of Australia, Germany, the United Kingdom of Great Britain and Northern Ireland, the Irish Free State, New Zealand and Norway, desiring to secure the prosperity of the whaling industry and, for that purpose, to maintain the stock of whales, have agreed as follows:

Article I

The contracting Governments will take appropriate measures to ensure the application of the provisions of the present Agreement and the punishment of infractions against the said provisions, and, in particular, will maintain at least one inspector of whaling on each factory ship under their jurisdiction. The inspectors shall be appointed and paid by Governments.

Article 2

The present Agreement applies to factory ships and whale catchers and to land stations as defined in Article 18 under the jurisdiction of the contracting Governments, and to all waters in which whaling is prosecuted by such factory ships and/or whale catchers.

Article 3

Prosecutions for infractions against or contraventions of the present Agreement and the regulations made thereunder shall be instituted by the Government or a Department of the Government.

Article 4

It is forbidden to take or kill Grey Whales and/or Right Whales.

Article 5

It is forbidden to take or kill any Blue, Fin, Humpback or Sperm whales below the following lengths, viz.:

(a) Blue whales	70 feet,
(b) Fin whales	55 feet,
(c) Humpback whales	35 feet,
(d) Sperm whales	35 feet.

Article 6

It is forbidden to take or kill calves, or suckling whales or female whales which are accompanied by calves or suckling whales.

Article 7

It is forbidden to use a factory ship or a whale catcher attached thereto for

the purpose of taking or treating baleen whales in any waters south of 40°
South Latitude, except during the period from the 8th day of December to
the 7th day of March following, both days inclusive, provided that in the
whaling season 1937–38 the period shall extend to the 15th day of March,
1938, inclusive.

Article 8

It is forbidden to use a land station or a whale catcher attached thereto for
the purpose of taking or treating whales in any area or in any waters for
more than six months in any period of twelve months, such period of six
months to be continuous.

Article 9

It is forbidden to use a factory ship or a whale catcher attached thereto
for the purpose of taking or treating baleen whales in any of the following
areas, viz.:

(a) In the Atlantic Ocean north of 40° South Latitude and in the
Davis Strait, Baffin Bay and Greenland Sea;

(b) In the Pacific Ocean east of 150° West Longitude between 40°
South Latitude and 35° North Latitude;

(c) In the Pacific Ocean west of 150° West Longitude between 40°
South Latitude and 20° North Latitude;

(d) In the Indian Ocean north of 40° South Latitude.

Article 10

Notwithstanding anything contained in this Agreement, any contracting
Government may grant to any of its nationals a special permit authorising that national to kill, take and treat whales for purposes of scientific
research subject to such restrictions as to number and subject to such other
conditions as the contracting Government thinks fit, and the killing,
taking and treating of whales in accordance with the terms in force under
this Article shall be exempt from the operation of this Agreement.

Any contracting Government may at any time revoke a permit granted
by it under this Article.

Article 11

The fullest possible use shall be made of all whales taken. Except in the case of whales or parts of whales intended for human food or for feeding animals, the oil shall be extracted by boiling or otherwise for all blubber, meat (except the meat of sperm whales) and bones other than the internal organs, whale bone and flippers, of all whales delivered to the factory ship or land station.

Article 12

There shall not at any time be taken for delivery to any factory ship or land station a greater number of whales than can be treated efficiently and in accordance with Article 11 of the present Agreement by the plant and personnel therein within a period of thirty-six hours from the time of the killing of each whale.

Article 13

Gunners and crews of factory ships, land stations and whale catchers shall be engaged on terms such that their remuneration shall depend to a considerable extent upon such factors as the species, size and yield of whales taken, and not merely upon the number of the whales taken, and no bonus or other remuneration, calculated by reference to the results of their work, shall be paid to the gunners and crews of whale catchers in respect of any whales the taking of which is forbidden by this Agreement.

Article 14

With a view to the enforcement of the preceding Article, each contracting Government shall obtain, in respect of every whale catcher under its jurisdiction, an account showing the total emolument of each gunner and member of the crew and the manner in which the emolument of each of them is calculated.

Article 15

Articles 5, 9, 13 and 14 of the present Agreement, in so far as they impose obligations not already in force, shall not until the 1st day of December, 1937, apply to factory ships, land stations or catchers attached thereto which are at present operating or which have already taken practical measures with a view to whaling operations during the period before the said date. In respect of such factory ships, land stations and whale catchers, the Agreement shall in any event come into force on the said date.

Article 16

The contracting Governments shall obtain with regard to all factory ships and land stations under their jurisdiction records of the number of whales of each species treated at each factory ship or land station and as to the aggregate amounts of oil of each grade and quantities of meal, guano and other products derived from them, together with particulars with respect to each whale treated in the factory ship or land station as to the date and place of taking, the species and sex of the whale, its length and, if it contains a foetus, the length and sex, if ascertainable, of the foetus.

Article 17

The contracting Governments shall, with regard to all whaling operations under their jurisdiction, communicate to the International Bureau for Whaling Statistics at Sandefjord in Norway the statistical information specified in Article 16 of the present Agreement together with any information which may be collected or obtained by them in regard to the calving grounds and migration routes of whales.

In communicating this information the Governments shall specify:

(a) The name and tonnage of each factory ship;

(b) The number and aggregate tonnage of the whale catchers;

(c) A list of the land stations which were in operation during the period concerned.

Article 18

In the present Agreement the following expressions have the meanings respectively assigned to them, that is to say:

"Factory ship" means a ship in which or on which whales are treated whether wholly or in part;

"Whale catcher" means a ship used for the purpose of hunting, taking, towing, holding on to, or scouting for whales;

"Land station" means a factory on the land, or in the territorial waters adjacent thereto, in which or at which whales are treated whether wholly or in part;

"Baleen whale" means any whale other than a toothed whale;

"Blue whale" means any whale known by the name of blue whale, Sibbald's rorqual or sulphur bottom;

"Fin whale" means any whale known by the name of common finback, common finner, common rorqual, finback, fin whale, herring whale, razorback, or true fin whale;

"Grey whale" means any whale known by the name of grey whale, California grey, devil fish, hard head, mussel digger, grey back, rip sack;

"Humpback whale" means any whale known by the name of bunch, humpback, humpback whale, humpbacked whale, hump whale or hunchbacked whale;

"Right whale" means any whale known by the name of Atlantic right whale, Arctic right whale, Biscayan right whale, bowhead, great polar whale, Greenland right whale, Greenland whale, Nordkaper, North Atlantic right whale, North Cape whale, Pacific right whale, pigmy right whale, Southern pigmy right whale or Southern right whale;

"Sperm whale" means any whale known by the name of sperm whale, spermacet whale, cachalot or pot whale;

"Length" in relation to any whale means the distance measured on the level in a straight line between the tip of the upper jaw and the notch between the flukes of the tail.

Article 19

The present Agreement shall be ratified and the instruments of ratification shall be deposited with the Government of the United Kingdom of Great Britain and Northern Ireland as soon as possible. It shall come into force upon the deposit of instruments of ratification by a majority of the signatory Governments, which shall include the Governments of the United Kingdom, Germany and Norway; and for any other Government not included in such majority on the date of the deposit of its instrument of ratification.

The Government of the United Kingdom will inform the other Governments of the date on which the Agreement thus comes into force and the date of any ratification received subsequently.

Article 20

The present Agreement shall come into force provisionally on the 1st day of July, 1937, to the extent to which the signatory Governments are respectively able to enforce it; provided that if any Government within two months of the signature of the Agreement informs the Government of the United Kingdom that it is unwilling to ratify it the provisional application of the Agreement in respect of that Government shall thereupon cease.

The Government of the United Kingdom will communicate the name of any Government which has signified that it is unwilling to ratify the Agreement to the other Governments, any of whom may within one month of such communication withdraw its ratification or accession or signify its unwillingness to ratify as the case may be, and the provisional application of the Agreement in respect of that Government shall thereupon cease. Any such withdrawal or communication shall be notified to the Government of the United Kingdom, by whom it will be transmitted to the other Governments.

Article 21

The present Agreement shall, subject to the preceding Article, remain in force until the 30th day of June, 1938, and thereafter if, before that date, a majority of the contracting Governments, which shall include the Governments of the United Kingdom, Germany and Norway, shall have agreed to extend its duration. In the event of such extension it shall remain in force until the contracting Governments agree to modify it, provided that any contracting Government may, at any time after the 30th day of June, 1938, by giving notice on or before the 1st day of January in any year to the Government of the United Kingdom (who on receipt of such notice shall at once communicate it to the other contracting Governments) withdraw from the Agreement, so that it shall cease to be in force in respect of that Government after the 30th day of June following, and that any other contracting Government may, by giving notice in the like manner within one month of the receipt of such communication, withdraw also from the Agreement, so that it shall cease to be in force respecting it after the same date.

Article 22

Any Government which has not signed the present Agreement may accede thereto at any time after it has come into force. Accession shall be effected by means of a notification in writing addressed to the Government of the United Kingdom and shall take effect immediately after the date of its receipt.

The Government of the United Kingdom will inform all the Governments which have signed or acceded to the present Agreement of all accessions received and the date of their receipt.

In faith whereof the undersigned, being duly authorised, have signed the present Agreement.

Done in London the 8th day of June, 1937, in a single copy, which shall remain deposited in the archives of the Government of the United Kingdom of Great Britain and Northern Ireland, by whom certified copies will be transmitted to all the other contracting Governments.

> *Union of South Africa*
> *United States of America*
> *Argentine Republic*
> *Commonwealth of Australia*
> *Germany*
> *United Kingdom of Great Britain and Northern Ireland*
> *Irish Free State*
> *New Zealand*
> *Norway*

Protocol Amending the International Agreement of June 8th, 1937, for the Regulation of Whaling. Signed in London on June 24, 1938 (The "1938 London Protocol")

The Governments of the Union of South Africa, the United States of America, the Argentine Republic, the Commonwealth of Australia, Canada, Eire, Germany, the United Kingdom of Great Britain and Northern Ireland, New Zealand and Norway, desiring to introduce certain amendments into the International Agreement for the Regulation of Whaling, signed in London on the 8th June, 1937 (hereinafter referred to as the Principal Agreement) in accordance with the provisions of Article 21 thereof, have agreed as follows:

Article I

With reference to the provisions of Articles 5 and 7 of the Principal Agreement, it is forbidden to use a factory ship or a whale catcher attached thereto for the purpose of taking or treating humpback whales in any waters south of 40° South Latitude during the period from the 1st October, 1938, to the 30th September, 1939.

Article 2

Notwithstanding the provisions of Article 7 of the Principal Agreement, it is forbidden to use a factory ship or a whale catcher attached thereto for the purpose of taking or treating baleen whales in the waters south of 40° South Latitude from 70° West Longitude westwards as far as 160° West Longitude for a period of two years from the 8th day of December, 1938.

Article 3

1. No factory ship which has been used for the purpose of treating baleen whales south of 40° South Latitude shall be used for that purpose elsewhere within a period of twelve months from the end of the open season prescribed in Article 7 of the Principal Agreement.

2. Only such factory ships as have operated during the year 1937 within the territorial waters of any signatory Government shall, after the signature of this Protocol, so operate, and any such ships so operating shall be treated as land stations and remain moored in territorial waters in one position during the season and shall operate for not more than six months in any period of twelve months, such period of six months to be continuous.

Article 4

To Article 5 of the Principal Agreement there shall be added the following:
    "Except that blue whales of not less than 65 feet, fin whales of not less than 50 feet and sperm whales of not less than 30 feet in length may be taken for delivery to land stations provided that the meat of such whales is to be used for local consumption as human or animal food."

Article 5

To Article 7 of the Principal Agreement there shall be added the following:
    "Notwithstanding the above prohibition of treatment during a close season, the treatment of whales which have been taken during the open season may be completed after the end of the open season."

Article 6

In Article 8 of the Principal Agreement the word "baleen" shall be inserted after the word "treating."

Article 7

For the areas specified in (a), (b), (c), and (d) of Article 9 of the Principal Agreement there shall be substituted the following areas, viz.:

(a) In the waters north of 66° North Latitude, except that from 150° East Longitude eastwards as far as 140° West Longitude the taking or killing of whales by such ship or catcher shall be permitted between 66° North Latitude and 72° North Latitude;

(b) In the Atlantic Ocean and its dependent waters north of 40° South Latitude;

(c) In the Pacific Ocean and its dependent waters east of 150° West Longitude between 40° South Latitude and 35° North Latitude;

(d) In the Pacific Ocean and its dependent waters west of 150° West Longitude between 40° South Latitude and 20° North Latitude;

(e) In the Indian Ocean and its dependent waters north of 40° South Latitude.

Article 8

For Article 12 of the Principal Agreement there shall be substituted the following, viz.:

The taking of whales for delivery to a factory ship shall be so regulated or restricted by the master or person in charge of the factory ship that no whale carcase shall remain in the sea for a longer period than 33 hours from the time of killing to the time when it is taken up on to the deck of the factory ship for treatment.

Article 9

The present Protocol shall come into force provisionally on the first day of July, 1938, to the extent to which the signatory Governments are respectively able to enforce it.

Article 10

(i) The present Protocol shall be ratified and the instruments of ratification shall be deposited with the Government of the United Kingdom of Great Britain and Northern Ireland as soon as possible.

(ii) It shall come into force definitively upon the deposit of the instruments of ratification by the Governments of the United Kingdom, Germany and Norway.

(iii) For any other Government which is a Party to the Principal Agreement, the present Protocol shall come into force on the date of the deposit of its instrument of ratification or notification of accession.

(iv) The Government of the United Kingdom will inform the other Governments of the date on which the Protocol comes into force and the date of any ratification or accession received subsequently.

Article 11

(i) The present Protocol shall be open to accession by any Government which has not signed it and which accedes to the principal Agreement before the definitive entry into force of the Protocol.

(ii) Accession shall be effected by means of a notification in writing addressed to the Government of the United Kingdom and shall take effect immediately after the date of its receipt.

(iii) The Government of the United Kingdom will inform all the Governments which have signed or acceded to the present Protocol of all accessions received and the date of their receipt.

Article 12

Any ratification of or accession to the Principal Agreement which may be deposited or notified after the date of definitive coming into force of the present Protocol shall be deemed to relate to the Principal Agreement as amended by the present Protocol.

In witness whereof the undersigned, duly authorised thereto, have signed the present Protocol.

Done in London the twenty-fourth day of June, 1938, in a single copy, which shall be deposited in the archives of the Government of the United

Kingdom of Great Britain and Northern Ireland, by whom certified copies shall be communicated to all the signatory Governments.

> *Union of South Africa*
> *United States of America*
> *Argentine Republic*
> *Commonwealth of Australia*
> *Canada*
> *Eire*
> *Germany*
> *United Kingdom of Great Britain and Northern Ireland*
> *New Zealand*
> *Norway*

International Convention for the Regulation of Whaling. Signed in Washington on December 2, 1946 (The "1946 Washington Convention")

The Governments whose duly authorized representatives have subscribed hereto,

Recognizing the interest of the nations of the world in safeguarding for future generations the great natural resources represented by the whale stocks;

Considering that the history of whaling has seen overfishing of one area after another and of one species of whale after another to such a degree that it essential to protect all species of whales from further overfishing;

Recognizing that the whale stocks are susceptible of natural increases if whaling is properly regulated, and that increases in the size of whale stocks will permit increases in the number of whales which may be captured without endangering these natural resources;

Recognizing that it is in the common interest to achieve the optimum level of whale stocks as rapidly as possible without causing wide-spread economic and nutritional distress;

Recognizing that in the course of achieving these objectives, whaling operations should be confined to those species best able to sustain exploitation in order to give an interval for recovery to certain species of whales now depleted in numbers;

Desiring to establish a system of international regulation for the whale fisheries to ensure proper and effective conservation and development of whale stocks on the basis of the principles embodied in the provisions of the International Agreement for the Regulation of Whaling signed in London on June 8, 1937 and the protocols to that Agreement signed in London on June 24, 1938 and November 26, 1945; and

Having decided to conclude a convention to provide for the proper conservation of whale stocks and thus make possible the orderly development of the whaling industry;

Have agreed as follows:

Article I

1. This Convention includes the Schedule attached thereto which forms an integral part thereof. All references to "Convention" shall be understood as including the said Schedule either in its present terms or as amended in accordance with the provisions of Article V.

2. The Convention applies to factory ships, land stations, and whale catchers under the jurisdiction of the Contracting Governments, and to all waters in which whaling is prosecuted by such factory ships, land stations, and whale catchers.

Article II

As used in this Convention

1. "factory ship" means a ship in which or on which whales are treated whether wholly or in part;

2. "land station" means a factory on the land at which whales are treated whether wholly or in part;

3. "whale catcher" means a ship used for the purpose of hunting, taking, towing, holding on to, or scouting for whales;

4. "Contracting Government" means any Government which has deposited an instrument of ratification or has given notice of adherence to this Convention.

Article III

1. The Contracting Governments agree to establish an International Whaling Commission, hereinafter referred to as the Commission, to be composed of one member from each Contracting Government. Each member shall have one vote and may be accompanied by one or more experts and advisers.

2. The Commission shall elect from its own members a Chairman and Vice Chairman and shall determine its own Rules of Procedure. Decisions of the Commission shall be taken by a simple majority of those members voting except that a three-fourths majority of those members voting shall be required for action in pursuance of Article V. The Rules of Procedure may provide for decisions otherwise than at meetings of the Commission.

3. The Commission may appoint its own Secretary and staff.

4. The Commission may set up, from among its own members and experts or advisers, such committees as it considers desirable to perform such functions as it may authorize.

5. The expenses of each member of the Commission and of his experts and advisers shall be determined and paid by his own Government.

6. Recognizing that specialized agencies related to the United Nations will be concerned with the conservation and development of whale fisheries and the products arising therefrom and desiring to avoid duplication of functions, the Contracting Governments will consult among themselves within two years after the coming into force of this Convention to decide whether the Commission shall be brought within the framework of a specialized agency related to the United Nations.

7. In the meantime the Government of the United Kingdom of Great Britain and Northern Ireland shall arrange, in consultation with the other Contracting Governments, to convene the first meeting of the Commission, and shall initiate the consultation referred to in paragraph 6 above.

8. Subsequent meetings of the Commission shall be convened as the Commission may determine.

Article IV

1. The Commission may either in collaboration with or through independent agencies of the Contracting Governments or other public or private agencies, establishments, or organizations, or independently

    (a) encourage, recommend, or if necessary, organize studies and investigations relating to whales and whaling;

    (b) collect and analyze statistical information concerning the current condition and trend of the whale stocks and the effects of whaling activities thereon;

    (c) study, appraise, and disseminate information concerning methods of maintaining and increasing the populations of whale stocks.

2. The Commission shall arrange for the publication of reports of its activities, and it may publish independently or in collaboration with the International Bureau for Whaling Statistics at Sandefjord in Norway and other organizations and agencies such reports as it deems appropriate, as well as statistical, scientific, and other pertinent information relating to whales and whaling.

Article V

1. The Commission may amend from time to time the provisions of the Schedule by adopting regulations with respect to the conservation and utilization of whale resources, fixing (a) protected and unprotected species; (b) open and closed seasons; (c) open and closed waters, including the designation of sanctuary areas; (d) size limits for each species; (e) time, methods, and intensity of whaling (including the maximum catch of whales to be taken in any one season); (f) types and specifications of gear and apparatus and appliances which may be used; (g) methods of measurement; and (h) catch returns and other statistical and biological records.

2. These amendments of the Schedule (a) shall be such as are necessary to carry out the objectives and purposes of this Convention and to provide for the conservation, development, and optimum utilization of the whale resources; (b) shall be based on scientific findings; (c) shall not involve restrictions on the number or nationality of factory ships or land stations, nor allocate specific quotas to

any factory ship or land station or to any group of factory ships or land stations; and (d) shall take into consideration the interests of the consumers of whale products and the whaling industry.

3. Each of such amendments shall become effective with respect to the Contracting Governments ninety days following notification of the amendment by the Commission to each of the Contracting Governments, except that (a) if any Government presents to the Commission objection to any amendment prior to the expiration of this ninety-day period, the amendment shall not become effective with respect to any of the Governments for an additional ninety days; (b) thereupon, any other Contracting Government may present objection to the amendment at any time prior to the expiration of the additional ninety-day period, or before the expiration of thirty days from the date of receipt of the last objection received during such additional ninety-day period, whichever date shall be the later; and (c) thereafter, the amendment shall become effective with respect to all Contracting Governments which have not presented objection but shall not become effective with respect to any Government which has so objected until such date as the objection is withdrawn. The Commission shall notify each Contracting Government immediately upon receipt of each objection and withdrawal and each Contracting Government shall acknowledge receipt of all notifications of amendments, objections, and withdrawals.

4. No amendments shall become effective before July 1, 1949.

Article VI

The Commission may from time to time make recommendations to any or all Contracting Governments on any matters which relate to whales or whaling and to the objectives and purposes of this Convention.

Article VII

The Contracting Governments shall ensure prompt transmission to the International Bureau for Whaling Statistics at Sandefjord in Norway, or to such other body as the Commission may designate, of notifications and statistical and other information required by this Convention in such form and manner as may be prescribed by the Commission.

Article VIII

1.   Notwithstanding anything contained in this Convention, any Contracting Government may grant to any of its nationals a special permit authorizing that national to kill, take, and treat whales for purposes of scientific research subject to such restrictions as to number and subject to such other conditions as the Contracting Government thinks fit, and the killing, taking, and treating of whales in accordance with the provisions of this Article shall be exempt from the operation of this Convention. Each Contracting Government shall report at once to the Commission all such authorizations which it has granted. Each Contracting Government may at any time revoke any such special permit which it has granted.

2.   Any whales taken under these special permits shall so far as practicable be processed and the proceeds shall be dealt with in accordance with directions issued by the Government by which the permit was granted.

3.   Each Contracting Government shall transmit to such body as may be designated by the Commission, in so far as practicable, and at intervals of not more than one year, scientific information available to that Government with respect to whales and whaling, including the results of research conducted pursuant to paragraph 1 of this Article and to Article IV.

4.   Recognizing that continuous collection and analysis of biological data in connection with the operations of factory ships and land stations are indispensable to sound and constructive management of the whale fisheries, the Contracting Governments will take all practicable measures to obtain such data.

Article IX

1.   Each Contracting Government shall take appropriate measures to ensure the application of the provisions of this Convention and the punishment of infractions against the said provisions in operations carried out by persons or by vessels under its jurisdiction.

2.   No bonus or other remuneration calculated with relation to the results of their work shall be paid to the gunners and crews of whale catchers in respect of any whales the taking of which is forbidden by this Convention.

3. Prosecution for infractions against or contraventions of this Convention shall be instituted by the Government having jurisdiction over the offense.

4. Each Contracting Government shall transmit to the Commission full details of each infraction of the provisions of this Convention by persons or vessels under the jurisdiction of that Government as reported by its inspectors. This information shall include a statement of measures taken for dealing with the infraction and of penalties imposed.

Article X

1. This Convention shall be ratified and the instruments of ratification shall be deposited with the Government of the United States of America.

2. Any Government which has not signed this Convention may adhere thereto after it enters into force by a notification in writing to the Government of the United States of America.

3. The Government of the United States of America shall inform all other signatory Governments and all adhering Governments of all ratifications deposited and adherences received.

4. The Convention shall, when instruments of ratification have been deposited by at least six signatory Governments, which shall include the Governments of the Netherlands, Norway, the Union of Soviet Socialist Republics, the United Kingdom of Great Britain and Northern Ireland, and the United States of America, enter into force with respect to those Governments and shall enter into force with respect to each Government which subsequently ratifies or adheres on the date of the deposit of its instrument of ratification or the receipt of its notification of adherence.

5. The provisions of the Schedule shall not apply prior to July 1, 1948. Amendments to the Schedule adopted pursuant to Article V shall not apply prior to July 1, 1949.

Article XI

Any Contracting Government may withdraw from this Convention on June thirtieth of any year by giving notice on or before January first of the same year to the depositary Government, which upon receipt of such a notice shall at once communicate it to the other Contracting Governments. Any other Contracting Government may, in like manner, within one month of the receipt of a copy of such a notice from the depositary Government, give notice of withdrawal, so that the Convention shall cease to be in force on June thirtieth of the same year with respect to the Government giving such notice of withdrawal.

This Convention shall bear the date on which it is opened for signature and shall remain open for signature for a period of fourteen days thereafter.

IN WITNESS WHEREOF the undersigned, being duly authorized, have signed this Convention.

DONE in Washington this second day of December 1946, in the English language, the original of which shall be deposited in the archives of the Government of the United States of America. The Government of the United States of America shall transmit certified copies thereof to all the other signatory and adhering Governments.

Argentina	*New Zealand*
Australia	*Norway*
Brazil	*Peru*
Canada	*Union of Soviet Socialist Republics*
Chile	*United Kingdom of Great Britain and Northern Ireland*
Denmark	*United States of America*
France	*Union of South Africa*
The Netherlands	

Schedule

1. (a) There shall be maintained on each factory ship at least two inspectors of whaling for the purpose of maintaining twenty-four hour inspection. These inspectors shall be appointed and paid by the Government having jurisdiction over the factory ship.

    (b) Adequate inspection shall be maintained at each land station. The inspectors serving at each land station shall be appointed and paid by the Government having jurisdiction over the land station.

2. It is forbidden to take or kill gray whales or right whales, except when the meat and products of such whales are to be used exclusively for local consumption by the aborigines.

3. It is forbidden to take or kill calves or suckling whales or female whales which are accompanied by calves or suckling whales.

4. It is forbidden to use a factory ship or a whale catcher attached thereto for the purpose of taking or treating baleen whales in any of the following areas:

(a) in the waters north of 66° North Latitude except that from 150° East Longitude eastward as far as 140° West Longitude the taking or killing of baleen whales by a factory ship or whale catcher shall be permitted between 66° North Latitude and 72° North Latitude;

(b) in the Atlantic Ocean and its dependent waters north of 40° South Latitude;

(c) in the Pacific Ocean and its dependent waters east of 150° West Longitude between 40° South Latitude and 35° North Latitude;

(d) in the Pacific Ocean and its dependent waters west of 150° West Longitude between 40° South Latitude and 20° North Latitude;

(e) in the Indian Ocean and its dependent waters north of 40° South Latitude.

5. It is forbidden to use a factory ship or a whale catcher attached thereto for the purpose of taking or treating baleen whales in the waters south of 40° South Latitude from 70° West Longitude westward as far as 160° West Longitude.

6. It is forbidden to use a factory ship or a whale catcher attached thereto for the purpose of taking or treating humpback whales in any waters south of 40° South Latitude.

7. (a) It is forbidden to use a factory ship or a whale catcher attached thereto for the purpose of taking or treating baleen whales in any waters south of 40° South Latitude, except during the period from December 15 to April 1 following, both days inclusive.

(b) Notwithstanding the above prohibition of treatment during a closed season, the treatment of whales which have been taken during the open season may be completed after the end of the open season.

8. (a) The number of baleen whales taken during the open season caught in any waters south of 40° South Latitude by whale catchers attached to factory ships under the jurisdiction of the Contracting Governments shall not exceed sixteen thousand blue-whale units.

(b) For the purposes of subparagraph (a) of this paragraph, blue-whale units shall be calculated on the basis that one blue whale equals:

(1) two fin whales or

(2) two and a half humpback whales or

(3) six sei whales.

(c) Notification shall be given in accordance with the provisions of Article VII of the Convention, within two days after the end of each calendar week, of data on the number of blue-whale units taken in any waters south of 40° South Latitude by all whale catchers attached to factory ships under the jurisdiction of each Contracting Government.

(d) If it should appear that the maximum catch of whales permitted by subparagraph (a) of this paragraph may be reached before April 1 of any year, the Commission, or such other body as the Commission may designate, shall determine, on the basis of the data provided, the date on which the maximum catch of whales shall be deemed to have been reached and shall notify each Contracting Government of that date not less than two weeks in advance thereof. The taking of baleen whales by whale catchers attached to factory ships shall be illegal in any waters south of 40° South Latitude after the date so determined.

(e) Notification shall be given in accordance with the provisions of Article VII of the Convention of each factory ship intending to engage in whaling operations in any waters south of 40° South Latitude.

9. It is forbidden to take or kill any blue, fin, sei, humpback, or sperm whales below the following lengths:

(a) blue whales	70 feet (21.3 meters)
(b) fin whales	55 feet (16.8 meters)
(c) sei whales	40 feet (12.2 meters)
(d) humpback whales	35 feet (10.7 meters)
(e) sperm whales	35 feet (10.7 meters)

except that blue whales of not less than 65 feet (19.8 meters), fin whales of not less than 50 feet (15.2 meters), and sei whales of not less than 35 feet (10.7 meters) in length may be taken for delivery to land stations provided that the meat of such whales is to be used for local consumption as human or animal food.

Whales must be measured when at rest on deck or platform, as accurately as possible by means of a steel tape measure fitted at the zero end with a spiked handle which can be stuck into the deck planking abreast of one end of the whale. The tape measure shall be stretched in a straight line parallel with the whale's body and read abreast the other end of the whale. The ends of the whale, for measurement purposes, shall be the point of the upper jaw and the notch between the tail flukes. Measurements, after being accurately read on the tape measure, shall be logged to the nearest foot: that is to say, any whale between 75'6" and 76'6" shall be logged as 76', and any whale between 76'6" and 77'6" shall be logged as 77'. The measurement of any whale which falls on an exact half foot shall be logged at the next half foot, e.g. 76'6" precisely, shall be logged as 77'.

10. It is forbidden to use a land station or a whale catcher attached thereto for the purpose of taking or treating baleen whales in any area or in any waters for more than six months in any period of twelve months, such period of six months to be continuous.

11. It is forbidden to use a factory ship, which has been used during a season in any waters south of 40° South Latitude for the purpose of treating baleen whales, in any other area for the same purpose within a period of one year from the termination of that season.

12. (a) All whales taken shall be delivered to the factory ship or land station and all parts of such whales shall be processed by boiling or otherwise, except the internal organs, whale bone and flippers of all whales, the meat of sperm whales and of parts of whales intended for human food or feeding animals.

    (b) Complete treatment of the carcasses of "Dauhval" and of whales used as fenders will not be required in cases where the meat or bone of such whales is in bad condition.

13. The taking of whales for delivery to a factory ship shall be so regulated or restricted by the master or person in charge of the factory ship that no whale carcass (except of a whale used as a fender) shall remain in the sea for a longer period than thirty-three hours from the time of killing to the time when it is taken up on to the deck of

the factory ship for treatment. All whale catchers engaged in taking whales must report by radio to the factory ship the time when each whale is caught.

14. Gunners and crews of factory ships, land stations, and whale catchers shall be engaged on such terms that their remuneration shall depend to a considerable extent upon such factors as the species, size, and yield of whales taken, and not merely upon the number of the whales taken. No bonus or other remuneration shall be paid to the gunners or crews of whale catchers in respect of the taking of milk-filled or lactating whales.

15. Copies of all official laws and regulations relating to whales and whaling and changes in such laws and regulations shall be transmitted to the Commission.

16. Notification shall be given in accordance with the provisions of Article VII of the Convention with regard to all factory ships and land stations of statistical information (a) concerning the number of whales of each species taken, the number thereof lost, and the number treated at each factory ship or land station, and (b) as to the aggregate amounts of oil of each grade and quantities of meal, fertilizer (guano), and other products derived from them, together with (c) particulars with respect to each whale treated in the factory ship or land station as to the date and approximate latitude and longitude of taking, the species and sex of the whale, its length and, if it contains a foetus, the length and sex, if ascertainable, of the foetus. The data referred to in (a) and (c) above shall be verified at the time of the tally and there shall also be notification to the Commission of any information which may be collected or obtained concerning the calving grounds and migration routes of whales.

In communicating this information there shall be specified:

(a) the name and gross tonnage of each factory ship;

(b) the number and aggregate gross tonnage of the whale catchers;

(c) a list of the land stations which were in operation during the period concerned.

17. Notwithstanding the definition of land station contained in Article II of the Convention, a factory ship operating under the jurisdiction of a Contracting Government, and the movements of which are con-

fined solely to the territorial waters of that Government, shall be subject to the regulations governing the operation of land stations within the following areas:

(a) on the coast of Madagascar and its dependencies, and on the west coasts of French Africa;

(b) on the west coast of Australia in the area known as Shark Bay and northward to Northwest Cape and including Exmouth Gulf and King George's Sound, including the port of Albany; and on the east coast of Australia, in Twofold Bay and Jervis Bay.

18.   The following expressions have the meanings respectively assigned to them, that is to say:

"baleen whale" means any whale other than a toothed whale;

"blue whale" means any whale known by the name of blue whale, Sibbald's rorqual, or sulphur bottom;

"fin whale" means any whale known by the name of common finback, common rorqual, finback, finner, fin whale, herring whale, razorback, or true fin whale;

"sei whale" means any whale known by the name of *Balaenoptera borealis,* sei whale, Rudolphi's rorqual, pollack whale, or coalfish whale, and shall be taken to include *Balaenoptera brydei,* Bryde's whale;

"gray whale" means any whale known by the name of gray whale, California gray, devil fish, hard head, mussel digger, gray back, rip sack;

"humpback whale" means any whale known by the name of bunch, humpback, humpback whale, hum[p]backed whale, hump whale, or hunchbacked whale;

"right whale" means any whale known by the name of Atlantic right whale, Arctic right whale, Biscayan right whale, bowhead, great polar whale, Greenland right whale, Greenland whale, Nordkaper, North Atlantic right whale, North Cape whale, Pacific right whale, pigmy right whale, Southern pigmy right whale, or Southern right whale;

"sperm whale" means any whale known by the name of sperm whale, spermacet whale, cachalot, or pot whale;

"Dauhval" means any unclaimed dead whale found floating.

Notes

Introduction

1. Aldo Leopold, *Game Management* (New York: Charles Scribner's Sons, 1933), 3, 21. On Aldo Leopold, see especially Curt Meine, *Aldo Leopold: His Life and Work* (Madison: University of Wisconsin Press, 1988); and Susan Flader and J. Baird Callicott, eds., *The River of the Mother of God and Other Essays* (Madison: University of Wisconsin Press, 1991).

2. Leopold, *Game Management*, 4–5.

3. The complete texts of nearly every international environmental treaty signed since 1750 is available in Bernd Rüster and Bruno Simma, eds., *International Protection of the Environment: Treaties and Related Documents* (Dobbs Ferry, NY: Oceana Publications, 1975); and Michael R. Molitor, ed., *International Environmental Law: Primary Materials* (Deventer, the Netherlands: Kluwer, 1991). The Rüster and Simma series covers the period up to 1975, Molitor the post-1975 period.

4. The fields of global nature conservation and environmental law are still in their infancy. Among the surveys that handle international animal protection are Robert Boardman, *International Organization and the Conservation of Nature* (Bloomington: Indiana University Press, 1981); Simon Lyster, *International Wildlife Law: An Analysis of International Treaties Concerned with the Conservation of Wildlife* (Cambridge: Grotius Publications, 1985); John M. MacKenzie, *The Empire of Nature* (Manchester, UK: University of Manchester Press, 1988); John McCormick, *Reclaiming Paradise: The Global Environmental Movement* (Bloomington: Indiana University Press, 1989); Philippe Sands, ed., *Greening International Law* (New York: New Press, 1994); and Kurkpatrick Dorsey, *The Dawn of Conservation Diplomacy: U.S.-Canadian Wildlife Protection Treaties in the Progressive Era* (Seattle: University of Washington Press, 1998).

5. David Day makes much the same point in *The Whale War* (London: Routledge and Kegan Paul, 1987), 29: "The International Whaling Commission (IWC)

was a 'whalers' club.' Like some exclusive big game shooting club, the members of the IWC since 1948 came together once a year before the opening of the killing season. They sat around a big table and smoked cigars, they had drinks in the bar and compared profits and talked of the good old days when the vast herds of great blue whales made life easy. Then they sat down at the table and bargained until they agreed among themselves what the sporting number of whales to be bagged this year would be. The quota was based on virtually non-existent science and a lot of wishful thinking. The hunt was a kind of 'gentlemen's agreement' between nations to abide by sporting rules: an exact date for the opening of the season, a ban on the killing of nursing mothers and undersized whales, and an immediate end to the killing when the quota was reached."

6. Cited by Alistair Graham, *The Gardeners of Eden* (London: George Allen and Unwin, 1973), 83. On poaching, see especially Charles Chenevix Trench, *The Poacher and the Squire: A History of Poaching and Game Preservation in England* (London: Longman, 1967); E. P. Thompson, *Whigs and Hunters: The Origin of the Black Act* (London: Allen Lane, 1975); Edward I. Steinhart, *Black Poachers, White Hunters: A Social History of Hunting in Colonial Kenya* (Athens: Ohio University Press, 2006); and Karl Jacoby, *Crimes against Nature: Squatters, Poachers, Thieves, and the Hidden History of American Conservation* (Berkeley: University of California Press, 2001).

7. Aldo Leopold, *Round River: From the Journals of Aldo Leopold*, ed. Luna B. Leopold (New York: Oxford University Press, 1953), 145–46.

Chapter 1: Africa's Apartheid Parks

1. On Hermann von Wissmann, see Bernhard Gissibl, "German Colonialism and the Beginnings of International Wildlife Preservation in Africa," *GHI-Bulletin*, Supplement 3 (2006): 121–43. Wissmann's official title was Reich commissioner from 1888 to 1891 and governor from 1895 to 1896, but this was essentially the same position under different names.

2. The literature on African colonialism is vast. Good starting points include Adu A. Boahen, *Africa under Colonial Domination 1880–1935* (Berkeley: University of California Press, 1985); and Thomas Pakenham, *The Scramble for Africa: White Man's Conquest of the Dark Continent from 1876 to 1912* (New York: Avon Books, 1991). See also Kevin Shillington, *History of Africa* (New York: St. Martin's Press, 1995); and Philip Curtin, Steven Feierman, Leonard Thompson, and Jan Vansina, *African History: From Earliest Times to Independence* (London: Longman, 1995).

3. On Belgian colonialism, see Adam Hochschild, *King Leopold's Ghost: A Story of Greed, Terror, and Heroism in Colonial Africa* (Boston: Houghton Mif-

flin, 1998). On German colonialism, see A. J. P. Taylor, *Germany's First Bid for Colonies, 1884–1885: A Move in Bismarck's Foreign Policy* (London: Macmillan, 1938); and Anna Perras, *Carl Peters and German Imperialism, 1856–1918: A Political Biography* (Oxford: Clarendon Press, 2004).

4. The best general survey of Africa's national parks is Nicholas Luard, *The Wildlife Parks of Africa* (London: Michael Joseph, 1985), though it is now somewhat dated. See also Richard Despard Estes, *The Behavior Guide to African Mammals, Including Hoofed Mammals, Carnivores, Primates* (Berkeley: University of California Press, 1991), xix, for a list of the major parks and reserves.

5. Bernard Grzimek and Michael Grzimek, *Serengeti Shall Not Die* (New York: E. P. Dutton, 1961), 309–12. Among other things, Grzimek undertook a sophisticated aerial count of the Serengeti animals. He tracked 16 mammal species and 1 bird (the ostrich) and concluded that there were a total of 366,980 animals in Serengeti (not the 1 million previously assumed). By far the most numerous were Thomson's and Grant's gazelles (194,654 together), the wildebeest (99,481), and Burchell's zebra (57,199). Like many others, Grzimek showed little sympathy for the rights of local Africans, and he undertook this survey in order to keep the park from being reduced in size at a time when the British government was under pressure from the Masai, who wanted to regain some of their lost grazing space.

6. On Nairobi Park, see Edward I. Steinhart, *Black Poachers, White Hunters: A Social History of Hunting in Colonial Kenya* (Athens: Ohio University Press, 2006), 190–91.

7. Roland Oliver and Michael Crowder, eds., *Cambridge Encyclopedia of Africa* (Cambridge: Cambridge University Press, 1981), 311.

8. The literature on African wildlife and conservation is vast, but a few works stand out: Philip D. Curtin, *The Image of Africa* (Madison: University of Wisconsin Press, 1973); David Anderson and Richard Grove, eds., *Conservation in Africa: Peoples, Policies and Practice* (Cambridge: Cambridge University Press, 1987); John M. MacKenzie, *The Empire of Nature* (Manchester, UK: University of Manchester Press, 1988); John M. MacKenzie, ed., *Imperialism and the Natural World* (Manchester, UK: Manchester University Press, 1990); Jonathan S. Adams and Thomas O. McShane, *The Myth of Wild Africa: Conservation without Illusion* (New York: W. W. Norton, 1992); Raymond Bonner, *At the Hand of Man: Peril and Hope for Africa's Wildlife* (New York: Alfred A. Knopf, 1993); Melissa Leach and Robin Mearns, eds., *The Lie of the Land: Challenging the Received Wisdom on the African Environment* (Portsmouth, NH: Heinemann, 1996); James McCann, *Green Land, Brown Land, Black Land* (Portsmouth, NH: Heinemann, 1999); and Terence Ranger, *Voices from the Rocks: Nature, Culture, and History in the Matopos Hills of Zimbabwe* (Oxford: James Currey, 1999).

For region-specific analyses, see especially Stuart Marks, *Imperial Lion: Human Dimensions of Wildlife Management in Central Africa* (Boulder, CO: Westview Press, 1984); Jane Carruthers, *The Kruger National Park: A Social and Political History* (Pietermaritzburg, South Africa: University of Natal Press, 1995); Gregory Maddox, James Giblin, and Isaria Kimambo, eds., *Custodians of the Land: Ecology & Culture in the History of Tanzania* (Athens: Ohio University Press, 1996); and Steinhart, *Black Poachers, White Hunters.*

9. Cited by Adams and McShane, *Myth of Wild Africa,* 25.

10. Cited by William M. Adams, *Against Extinction: The Story of Conservation* (London: Earthscan, 2004), 105.

11. Cited by Roderick Frazier Nash, *Wilderness and the American Mind* (New Haven, CT: Yale University Press, 2001 [1967]), 365.

12. See especially H. A. Bryden, ed., *Great and Small Game of Africa: An Account of the Distribution, Habits, and Natural History of the Sporting Mammals, with Personal Hunting Experiences* (London: Rowland Ward, 1899); and Richard Tjader, *The Big Game of Africa* (New York: D. Appleton, 1910). Bryden's book served as the principle reference work for the delegates at the 1900 London Conference. For an excellent historical survey of the colonial hunting ethos, see Mackenzie, *Empire of Nature.* He focuses on Great Britain, but his analysis can be extended to other imperial powers as well.

13. Estes, *Behavior Guide to African Mammals,* 271–77 and 449–56.

14. Ibid., 3–6 and 209.

15. George Frederick Kunz, *Ivory and the Elephant in Art, in Archaeology and in Science* (Garden City, NY: Doubleday, 1916), 437.

16. Estes, *Behavior Guide to African Mammals,* 259–67.

17. Derek Wilson and Peter Ayerst, *White Gold: The Story of African Ivory* (London: Heinemann, 1967), 10–12.

18. See especially E. D. Moore, *Ivory: Scourge of Africa* (New York: Harper and Brothers, 1931), 40–46. For a sympathetic portrayal of traditional techniques, see Steinhart, *Black Poachers, White Hunters,* esp. 28–29. For a critical perspective, see Julian Huxley, *The Conservation of Wild Life and Natural Habitats in Central and East Africa: Report on a Mission Accomplished for UNESCO, July–September 1960* (Paris: UNESCO, 1961), 17–19.

19. Moore, *Ivory,* 18–36 and 47–62. For a more detailed analysis of the ivory-and-slave trade in East Africa, see Edward A. Alpers, *Ivory and Slaves: Changing Patterns of International Trade in East Central Africa to the Later Nineteenth Century* (Berkeley: University of California Press, 1975), 234–53; Moses D. E. Nwulia, *Britain and Slavery in East Africa* (Washington, DC: Three Continents Press, 1975), 125–204; and R. W. Beachey, *The Slave Trade of Eastern Africa* (New York: Barnes and Noble, 1976), 181–262. For an overview on Zanzibar, see

Abdul Sheriff, *Slaves, Spices, and Ivory in Zanzibar* (Athens: Ohio University Press, 1987); and Erik Gilbert, *Dhows and the Colonial Economy of Zanzibar, 1860–1970* (Athens: Ohio University Press, 2004).

20. Kunz, *Ivory and the Elephant,* 443; and Wilson and Ayerst, *White Gold,* 154–56.

21. Henry Drummond, *Tropical Africa* (London: Hodder and Stoughton, 1889), 19.

22. Cited by Moore, *Ivory,* 172.

23. H. A. Bryden, "The Extermination of Game in South Africa," *Fortnightly Review* 62 (1894): 540; and MacKenzie, *Empire of Nature,* 87.

24. On South African game preservation, see John A. Pringle, *The Conservationists and the Killers: The Story of Game Protection and the Wildlife Society of Southern Africa* (Cape Town: T. V. Bulpin, 1982); Jane Carruthers, "Game Protection in the Transvaal 1846–1926" (PhD diss., University of Cape Town, 1988); and Carruthers, *Kruger National Park.*

25. J. Stevenson-Hamilton, *South African Eden: From Sabi Game Reserve to Kruger National Park* (London: Cassell, 1937), xviii.

26. John M. MacKenzie, "Chivalry, Social Darwinism and Ritualised Killing: The Hunting Ethos in Central Africa up to 1914," in Anderson and Grove, *Conservation in Africa,* 41–62. MacKenzie analyzes safari hunting more fully in *Empire of Nature.* See also Harriet Ritvo, "Destroyers and Preservers: Big Game in the Victorian Empire," *History Today* 52, no. 1 (January 2002): 33–39.

27. Tjader, *Big Game of Africa,* 23.

28. Ibid., 299.

29. Cited by MacKenzie, *Empire of Nature,* 157–58.

30. Richard Waller, "Emutai: Crisis and Response in Maasailand 1883–1902," in *The Ecology of Survival: Case Studies from Northeast African History,* ed. Douglas H. Johnson and David M. Anderson, 73–112 (London: Lester Crook, 1988). See also Juhani Koponen, "War, Famine, and Pestilence in Late Precolonial Tanzania: A Case for a Heightened Mortality," *International Journal of African Historical Studies* 21, no. 4 (1988): 637–76, for an excellent analysis of disease patterns in East Africa.

31. Nora Kelly, "In Wildest Africa: The Preservation of Game in Kenya 1895–1933" (PhD diss., Simon Fraser University, 1978), 208–9.

32. Andrew R. Carlson, *German Foreign Policy, 1890–1914, and Colonial Policy to 1914: A Handbook and Annotated Bibliography* (Metuchen, NJ: Scarecrow Press, 1970), 57.

33. Letter from Governor Hesketh Bell to Earl of Elgin, January 21, 1908, in *Further Correspondence Relating to the Preservation of Wild Animals in Africa,* Cd. 4472 (January 1909), 81.

34. Letter from Major von Wissmann to Baron Richtofen, April 2, 1897, in *Correspondence Relating to the Preservation of Wild Animals in Africa*, Cd. 3189 (November 1906), 34–35.

35. See Gissibl, "German Colonialism," 126–28.

36. Letter from Marquess of Salisbury to Hardinge (East African Protectorate) and Berkeley (Uganda Protectorate), May 27, 1896, in *Correspondence Relating to the Preservation of Wild Animals in Africa*, 1.

37. Cited by Gissibl, "German Colonialism," 129.

38. Letter from Sir F. Lascelles to the Marquess of Salisbury, January 2, 1897, in *Further Correspondence Respecting East Africa*, Part 48/6951 (1897), 4.

39. Memorandum by Mr. S. L. Hinde of the East Africa Protectorate, April 1900, in *Correspondence Relating to the Preservation of Wild Animals in Africa*, 100–101.

40. Letter from Arthur Hardinge to Marquess of Salisbury, February 19, 1900, in *Further Correspondence Respecting East Africa*, Part 60/7404 (1900), 119.

41. Kayser's statements are reported by Martin Gosselin to Marquess of Salisbury, July 15, 1896, in *Further Correspondence Respecting East Africa*, Part 46/6861 (1897), 83. On the tusk-size issue, see also the comments by Acting Commissioner Sharpe of the British Central Africa Protectorate (Nyasaland) to the Marquess of Salisbury, September 9, 1896, *Correspondence Relating to the Preservation of Wild Animals in Africa*, 27. Sharpe stated: "My own opinion is, and has always been, that there is only one method by which we can stop this slaughter of elephants of all sizes, and that is that all the Powers who hold territory in Africa should agree to prohibit the export of tusks of less weighing than, say, 15 lbs. each. If this course were taken by all the Powers, as soon as it became known throughout tropical Africa (and that would probably take two or three years) the killing of small elephants would cease, as the ivory would be no longer of value to the natives, and they would not waste their powder on anything but the larger bulls. For one Power alone, or two or three, however, to pass such a Regulation as this would be useless unless all the others joined in it, as it would simply result in the ivory of small size no longer being exported through those particular territories where it was forbidden, but going by new channels to the territories which had no such Regulations."

42. Letter from Götzen to C. Eliot, August 19, 1903, in *Correspondence Relating to the Preservation of Wild Animals in Africa*, 210–11. Several years later, when the British contemplated raising the tusk-size minimum to twenty-five pounds, Commissioner Sharpe warned the Earl of Elgin: "Experience has shown that where a prohibition is made in a British African possession against the export or import of any article of trade—unless a similar prohibition is made *and actually observed* in neighbouring territories belonging to other Powers—

there is a resulting loss of trade to British merchants and gain of trade by those carrying on operations in these adjoining territories. This was notably the case here in connection with the prohibition against the sale of trade gun-powder under existing international conventions. The sale of this article ceased in British Central Africa many years ago, but remained in full operation in the Congo Free State and German and Portuguese territories, the result being that ivory and other local products formerly sold within this Protectorate subsequently found their way to a great extent to these neighbouring territories where powder could be obtained. The second instance was when the prohibition against the export of tusks under 11 lbs. came into force. The law was strictly carried out in this Protectorate, but not in adjoining territories, with the result that many small tusks found their way to Portuguese and German territory, and were readily bought up by merchants there. Provided that Germany, the Congo Free State and Portugal all agree to raise the weight of exportable tusks to 25 lbs., and rigidly carry out the prohibition, then I am in entire concurrence with the proposal; but if the prohibition is to be limited to this Protectorate (and other Protectorates) it would undoubtedly operate hard on merchants here." Letter from Commissioner Sir Alfred Sharpe to the Earl of Elgin, October 17, 1906, in *Further Correspondence Relating to the Preservation of Wild Animals in Africa,* Cd. 4472, 25.

43. Three documents are central to the discussion over the convention: (1) the joint British-German Draft of Suggested Bases for Deliberations of an International Conference for the Protection of Wild Animals, Birds, and Fishes in Africa (the British-German Draft); (2) the British Foreign Office's Avant-Projet d'Acte Général (the Avant-Projet); and (3) the London Convention text itself, signed on May 19, 1900. The secondary literature on this treaty is sparse. See especially MacKenzie, *Empire of Nature,* 201–24; and Sherman Strong Hayden, *The International Protection of Wild Life* (New York: Columbia University Press, 1942), 36–42.

44. The southern zone was discussed several times. See especially the dispatch from the Colonial Office to the Foreign Office, March 13, 1899, in *Further Correspondence Respecting East Africa,* Part 56/7400 (1900), 261–62; and "Procès-Verbal de la Première Séance tenue le 25 Avril 1900," in *Further Correspondence Relating to the Preservation of Wild Animals in Africa,* Part 2/7822 (1902), 137–39.

45. The conference participants belonged to a civilization that considered the guillotine more "humane" than hanging, so it can be surmised that the length of time it took to die (rather than the killing itself) also perturbed them in regard to the African hunting methods. For a thorough discussion of African and European hunting practices, see MacKenzie, *Empire of Nature,* especially 54–84 and 204–10.

46. Foreign Office to Colonial Office, September 8, 1897, *Correspondence Relating to the Preservation of Wild Animals in Africa*, 44. The conference participants reaffirmed the efficacy of the Brussels Act over and over again in their discussions, but they left it out of the treaty text to avoid redundancy.

47. "Procès-Verbal de la Cinquième Séance tenue le 1er Mai 1900," in *Further Correspondence Relating to the Preservation of Wild Animals in Africa*, Part 2/7822, 154. See also Lankester's letter to the Foreign Office, June 24, 1901, in *Correspondence Relating to the Preservation of Wild Animals in Africa*, 145: "I would wish, first of all, to draw your attention to the powers given by Article III of the 'dispositions' adopted by the Conference of Plenipotentiaries on the Preservation of African Wild Animals, 1st May, 1900. The final clause of that Article was inserted on my suggestion with a view to such a case as that reported by Mr. [Val] Gielgud [of British South Africa Company], and gives power to dispense with the principles agreed upon 'dan un intérêt supérieur d'administration.' It is therefore within the provisions of the Agreement signed by the Plenipotentiaries for the Government to authorize the British South Africa Company to destroy buffalo, in order to protect domesticated cattle from disease."

48. "Draft of Suggested Bases for Deliberations of an International Conference for the Protection of Wild Animals, Birds, and Fishes in Africa," in *Correspondence Relating to the Preservation of Wild Animals in Africa*, 57. See also the dispatch from the Colonial Office to the Foreign Office, March 13, 1899, in *Further Correspondence Respecting East Africa*, Part 56/7400 (1900), 261–62.

49. The "Avant-Projet d'Acte Général," is included in the "Annexe au Procès-Verbal de la Séance de la Commission tenue le 25 Avril 1900," in *Further Correspondence Relating to the Preservation of Wild Animals in Africa*, Part 2/7822, 138–39.

50. Letter from Administrator R. B. Llewelyn (Gambia) to Mr. Chamberlain, March 21, 1899, *Correspondence Relating to the Preservation of Wild Animals in Africa*, 57–58.

51. "Procès-Verbal de la Cinquième Séance tenue le 1er Mai 1900," in *Further Correspondence Relating to the Preservation of Wild Animals in Africa*, Part 2/7822, 153.

52. These were the Earl of Hopetoun's words, in his opening address as conference president. "Procès-Verbal, Séance du 24 Avril 1900," ibid., 13.

53. "Procès-Verbal de la Quatrième Séance tenue le 30 Avril 1900," ibid., 151.

54. "Procès-Verbal de la Troisième Séance tenue le 27 Avril 1900," ibid., 147.

55. Ibid., 148.

56. "Procès-Verbal de la Deuxième Séance tenue le 26 Avril 1900," ibid., 141.

57. "Procès-Verbal de la Quatrième Séance tenue le 30 Avril 1900," ibid., 150.

58. "Procès-Verbal, Séance du 18 Mai 1900," ibid., 171–72.

59. Letter from Consul C. F. Cromie to Marquess of Landsdowne, September 14, 1902, in *Further Correspondence Respecting the Preservation of Wild Game in Africa*, Part 3/8384 (1904), 39.

60. Portugal and France made their reservations at the last working session of the conference. The Portuguese representative found fault with the southern demarcation line as outlined in Article I: "The particular situation of the Portuguese Colonies means that such a state of things would be more inconvenient for them than for any other of the Powers represented at the Conference. The French territories and those of the Congo are at the extreme north of the zone covered by Article I. They do not have neighbors who could exploit, in their own territories, or distribute freely from their ports, the products that would be hindered or prohibited elsewhere by an international agreement. The Portuguese Government has therefore charged its Plenipotentiary with declaring to the Conference that he cannot sign the Convention or ratify it unless Article I be modified in such a way as to include all the countries of southern Africa under its proposed restrictions." Then Léon Geoffray spoke for France in regard to the northern demarcation line: "In the opinion of the French Government, the measures adopted by the Conference cannot be effective unless all the countries included in the zone defined by Article I give their support to the Convention. The French Plenipotentiaries are ready to give their signatures to the Act if it is reformulated, but they warn the Conference that France will ratify it only after having been advised of the adherence of the countries which are part of the region, but which are not represented at the Conference, notably the Republic of Liberia and Abyssinia." "Procès-Verbal, Séance du 16 Mai 1900," in *Further Correspondence Relating to the Preservation of Wild Animals in Africa*, Part 2/7822, 168–69. Other documents concerning ratification are in *Further Correspondence Respecting the Preservation of Wild Game in Africa*, Part 3/8384; and *Further Correspondence Respecting the Preservation of Wild Game in Africa*, Part 4/8991 (1905).

61. The texts of the new game laws can be found in various parts of *Further Correspondence Relating to the Preservation of Wild Animals in Africa*, Cd. 4472. Many of the ordinances refer specifically to the stipulations in the London Convention. For a comprehensive report on the game reserves, see C. W. Hobley, "The London Convention of 1900," *Journal of the Society for the Preservation of the Fauna of the Empire* NS Part 20 (August 1933): 33–49.

62. For a history of the Fauna Society, see Richard Fitter, *The Penitent Butchers* (London: Collins, 1978); and Adams, *Against Extinction*. Edward North Buxton founded the society in 1903 to combat plans in Sudan to abandon the game reserve

north of the Sobat River and substitute it with an inferior one to the south. The society removed the "Wild" from its name after World War I and became the Fauna Preservation Society still later. It is now called Fauna and Flora International. It published the *Journal of the Society for the Preservation of the (Wild) Fauna of the Empire,* now called *Oryx.* The Fauna Society members denied being a group of "men who, having in earlier days taken their fill of big-game slaughter and the delights of the chase in wild, outlying parts of the earth, now, being smitten with remorse, and having reached a less strenuous term of life, think to condone our earlier bloodthirstiness by advocating the preservation of what we formerly chased or killed" (Fitter, *Penitent Butchers,* 9).

63. The parks and reserves are all listed country by country in American Committee for International Wild Life Protection, *African Game Protection: An Outline of the Existing Game Reserves and National Parks of Africa* (Cambridge, MA: American Committee for International Wild Life Protection, 1933). The list was updated in American Committee for International Wild Life Protection, *The London Convention for the Protection of African Fauna and Flora with Map and Notes on Existing African Parks and Reserves* (Cambridge, MA: American Committee for International Wild Life Protection, 1935).

64. Figures are from Nora Kelly, "In Wildest Africa," 138. Kelly explores in detail the contradiction between the goal of game protection and the goal of using game for generating revenue.

65. Quoted in Steinhart, *Black Poachers, White Hunters,* 152. See also Alistair Graham, *The Gardeners of Eden* (London: George Allen and Unwin, 1973), 56.

66. "Illegal Killing of Elephants and Rhinoceros, with Special Reference to the Italian Boundary of Kenya: Report of the Kenya Game Dept," PRO FO 371/11948 (1927). For a complete analysis of British-Italian problems over ivory and horns, see A. T. A. Ritchie's report, "Ivory Trade and Export to Italian Somaliland (April 14, 1931)," PRO FO 371/15395.

67. For Caldwell's comments, see PRO FO 371/12724 (1928) (italics added).

68. Peter T. Dalleo, "The Somali Role in Organized Poaching in Northeastern Kenya, c. 1909–1939," *International Journal of African Historical Studies* 12, no. 3 (1979): 472–82. See also Steinhart, *Black Poachers, White Hunters,* 174–77. Steinhart argues that the British-Italian negotiations laid the cornerstone for the second London Conference of 1933.

69. Letter from Commissioner Sadler (Uganda) to the Marquess of Lansdowne, May 1, 1903, *Correspondence Relating to the Preservation of Wild Animals in Africa,* 191–92. Sadler was relaying a report from one of his colleagues. The title "commissioner" was changed to "governor" a short time later.

70. His remarks are contained in a letter from the Governor of Uganda to the Secretary of State, April 22, 1909, in *Further Correspondence Relating to*

the *Preservation of Wild Animals in Africa,* Cd. 5136 (June 1910), 44. He added: "The elephants seem to have become even more bold than they were two years ago, and I came across numbers of flourishing gardens and plantations that had been absolutely wiped out by the herds of elephants that are roaming through the country. The complete destruction worked by these beasts is hardly credible, and the natives are getting desperate. Although a fair number of bulls are killed each year by sportsmen for the sake of their ivory, females and immature animals have, for many years past, enjoyed absolute immunity from danger. It is no uncommon thing to come across herds of between 100 and 300 cows, calves, and small bulls, and they are evidently increasing to such a degree that they are rendering uninhabitable a very large area of fertile and valuable land."

71. It was not until the 1950s that researchers began to comprehend that the elephant overpopulation problem was actually a subset of the habitat-compression problem. In 1900, elephants roamed across 70 percent of Uganda, but by 1970, the elephants' domain had shrunk to less than 20 percent. Elephants were "problematic" not because there were more of them in Uganda by 1970 (in fact there were fewer) but because they were being squeezed into less and less territory, which they tended to destroy because it was too small to meet their roaming and feeding needs. As cultivation expanded, so did the complaints about elephant rampages; as the complaints mounted, so did the culling campaigns; and as the culling campaigns increased, the elephant population spiraled downward—all under the "watchful" eye of a game department that loudly proclaimed its commitment to elephant protection. The classic study on Uganda elephants was done by R. M. Laws, I. S. C. Parker, and R. C. B. Johnstone, *Elephants and Their Habitats: The Ecology of Elephants in North Bunyoro, Uganda* (Oxford: Clarendon Press, 1975). For a similar investigation of Kenya's Tsavo Park, see I. S. C. Parker and Mohamed Amin, *Ivory Crisis* (London: Chatto and Windus, 1983).

72. On the disease, see John Ford, "African Trypanosomiasis: An Assessment of the Tsetse-Fly Problem Today," in *African Environment: Problems and Perspectives,* ed. Paul Richards, 67–72 (London: International African Institute, 1975); John Ford, *The Role of the Trypanosomiases in African Ecology: A Study of the Tsetse Fly Problem* (Oxford: Clarendon Press, 1971); and James Giblin, "Trypanosomiasis Control in African History: An Evaded Issue?" *Journal of African History* 31, no. 1 (1990): 59–80. On the impact of the disease on African society, see Helge Kjekshus, *Ecology Control and Economic Development in East African History: The Case of Tanganyika, 1850–1950* (London: Heinemann, 1977); Leroy Vail, "Ecology and History: The Example of Eastern Zambia," *Journal of Southern African Studies* 3, no. 2 (1977): 129–55; John McCracken, "Colonialism, Capitalism and the Ecological Crisis in Malawi: A Reassessment," in

Conservation in Africa: People, Policies and Practice, ed. David Anderson and Richard Grove, 63–77 (Cambridge: Cambridge University Press, 1987), 63–77; and Kirk Arden Hoppe, *Lords of the Fly: Sleeping Sickness Control in British East Africa, 1900–1960* (Westport, CT: Praeger, 2003).

73. On the tsetse-management debates, see especially Hoppe, *Lords of the Fly.* See also Thomas P. Ofcansky, "A History of Game Preservation in British East Africa, 1895–1963" (PhD diss., West Virginia University, 1981), 112–63; and Thomas P. Ofcansky, *Paradise Lost: A History of Game Preservation in East Africa* (Morgantown: West Virginia University Press, 2002), 47–63.

74. Letter from Rev. Dr. George Prentice to Acting Governor of Nyasaland, October 21, 1910, in *Further Correspondence Relating to the Preservation of Wild Animals in Africa,* Cd. 5775 (July 1911), 18–19 (italics in original).

75. Cited in John M. MacKenzie, "Experts and Amateurs: Tsetse, Nagana and Sleeping Sickness in East and Central Africa," in MacKenzie, *Imperialism and the Natural World,* 201.

76. Roben Mutwira, "Southern Rhodesian Wildlife Policy (1890–1953): A Question of Condoning Game Slaughter?" *Journal of Southern African Studies* 15, no. 2 (January 1989): 250–62; and MacKenzie, *Empire of Nature,* 226–56.

77. Buxton's remarks are in "Minutes of Proceedings at a Deputation from the Society for the Preservation of the Fauna of the Empire to the Right Honourable Alfred Lyttelton (His Majesty's Secretary for the Colonies)," Colonial Office, February 2, 1905, *Correspondence Relating to the Preservation of Wild Animals in Africa,* 252.

78. Letter from E. N. Buxton (of Fauna Society) to Colonial Office, December 30, 1909, *Further Correspondence Relating to the Preservation of Wild Animals in Africa,* Cd. 5136, 74.

79. Carruthers, "Game Protection in the Transvaal," 112–23. For a similar analysis of a reserve in Natal, see P. M. Brooks and I. A. W. Macdonald, "The Hluhluwe-Umfolozi Reserve: An Ecological Case History," in *Management of Large Mammals in African Conservation Areas: Proceedings of a Symposium Held in Pretoria, South Africa, 29–30 April 1982,* ed. R. Norman Owen-Smith, 51–77 (Pretoria, South Africa: HAUM, 1983).

80. The difficulties are detailed in Stevenson-Hamilton, *South African Eden.* The name was changed from Sabi to Kruger in order to win over reluctant members of the Volksraad. Stevenson-Hamilton commented on the change: "The man who *really* was responsible was R. K. Loveday . . . but the 'Kruger stunt' is I think a priceless value to us, and I would not for the world do aught but whisper otherwise. . . . I wonder what the old man, who *never in his life* thought of wild animals except as biltong, and who, with the idea that it did not matter much one way or the other, and in any case would not affect any

one except the town sportsmen, gave way under strong pressure exercised by Loveday and one or two others and allowed the reserve to be declared. I wonder, I repeat, what he would say could he see himself depicted as the '*Saviour of the South African game!!!*'" Cited in Carruthers, "Game Protection in the Transvaal," 364. See also Jane Carruthers, "Creating a National Park, 1910 to 1926," *Journal of Southern African Studies* 15, no. 2 (January 1989): 188–216.

81. Cited by Ofcansky, *Paradise Lost*, 83.

82. R. W. G. Hingston, "Proposed National Parks for Africa: A Paper Read at the Evening Meeting of the Society on 9 March 1931," *Geographical Journal* 77, no. 5 (May 1931): 402–6.

83. Ibid.

84. Ibid., 406.

85. Cited in MacKenzie, *Empire of Nature*, 81.

86. Economic Advisory Council, "Preparatory Committee for the International Conference for the Protection of the Fauna and Flora of Africa, 1933, Draft Second Report, February 21, 1933," PRO FO 371/16999 186008 Africa General (1933). The secondary literature on the 1933 treaty is small. See especially MacKenzie, *Empire of Nature*, 261–94; Simon Lyster, *International Wildlife Law: An Analysis of International Treaties Concerned with the Conservation of Wildlife* (Cambridge: Grotius Publications, 1985), 112–28; and Hayden, *International Protection of Wild Life*, 42–62.

87. Economic Advisory Council, "Preparatory Committee for the International Conference."

88. Ibid.

89. "Convention Relative to the Preservation of Fauna and Flora in the Natural State. Signed at London, November 8th, 1933," in Bernd Rüster and Bruno Simma, eds., *International Protection of the Environment: Treaties and Related Documents*, vol. 4 (Dobbs Ferry, NY: Oceana Publications, 1975), 1693.

90. Ibid., 1695.

91. Ibid., 1695–96.

92. Ibid., 1696–98.

93. Ibid., 1698–99.

94. Ibid., 1699–1700.

95. The conference did not keep minutes to its meetings but instead summarized the discussion in the Final Act. To identify the main areas of contention, I compared the text of the Draft Second Report with the text of the 1933 London Convention.

96. "Convention Relative to the Preservation of Fauna and Flora in the Natural State. Signed at London, November 8th, 1933," in Rüster and Simma, *International Protection of the Environment*, 1694.

97. "Economic Advisory Council, Fauna and Flora of Asia Committee (July 12th 1934)," PRO FO 371/18164 (1934). See also S. Harcourt-Smith, "The Applicability to Non-African Countries of the Agreement Concluded at the International Conference for the Protection of the Fauna and Flora of Africa (Sept. 6, 1934)," PRO FO 371/18164.

98. "Economic Advisory Council, Committee for the Protection of the Fauna and Flora of Asia, Australia and New Zealand (July 6th, 1939)," PRO FO 371/23544. The following governments were invited to attend the November 1939 conference: Anglo-Egyptian Sudan, Australia, Belgium, China, Egypt, France, Italy, Liberia, Nepal, Netherlands, New Zealand, Portugal, Siam, Southern Rhodesia, Spain, Union of South Africa, and the United States.

Chapter 2: The North American Bird War

1. William T. Hornaday, *Our Vanishing Wild Life: Its Extermination and Preservation* (New York: New York Zoological Society, 1913), 203. For a succinct biographical sketch of Hornaday's activities in the years between 1900 and 1930, see especially Frank Graham Jr., *Man's Dominion: The Story of Conservation in America* (Philadelphia: M. Evans, 1971), 179–224; and Gregory J. Dehler, "An American Crusader: William Temple Hornaday and Wildlife Protection in America, 1840–1940" (PhD diss., Lehigh University, 2001).

2. Hornaday, *Our Vanishing Wild Life*, 54–72 and 94–142.

3. Ibid. (quote on 54, statistic on 59).

4. Ibid., 247–57 (quotes on 247).

5. John C. Phillips, *Migratory Bird Protection in North America: The History of Control by the United States Federal Government and a Sketch of the Treaty with Great Britain* (Cambridge, MA: American Committee for International Wild Life Protection, 1934), 5.

6. Hornaday, *Our Vanishing Wild Life*, 64.

7. Ibid., 144–54 (quote on 146).

8. The demise of the passenger pigeon is well chronicled, most recently by Jennifer Price, *Flight Maps: Adventures with Nature in Modern America* (New York: Basic Books, 1999), 1–58.

9. Hornaday, *Our Vanishing Wild Life*, 17–33 (quote on 30; italics in original).

10. Price, *Flight Maps*, 57.

11. Graham, *Man's Dominion*, 23, 210–11.

12. Hornaday, *Our Vanishing Wild Life*, 5, 68–69, 148–49.

13. Helen Ossa, *They Saved Our Birds: The Battle Won and the War to Win* (New York: Hippocrene Books, 1973), 17–38.

14. Robin W. Doughty, *Feather Fashions and Bird Preservation* (Berkeley: University of California Press, 1975), 20–22.

15. Ibid., 153.

16. Price, *Flight Maps*, 57–109.

17. See William Haskell, *The American Game Protective and Propagation Association: A History* (New York, 1937), John F. Reiger, *American Sportsmen and the Origins of Conservation* (Corvallis: Oregon State University Press, 2001), and James B. Trefethen, *An American Crusade for Wildlife* (New York: Winchester Press, 1975) for a defense of hunters as conservationists. For a more critical perspective, see Samuel P. Hays, *Conservation and the Gospel of Efficiency: The Progressive Conservation Movement, 1890–1920* (Cambridge, MA: Harvard University Press, 1959), and Thomas R. Dunlap, *Saving America's Wildlife* (Princeton, NJ: Princeton University Press, 1988).

18. Robert H. Connery, *Governmental Problems in Wild Life Conservation* (New York: Columbia University Press, 1935), 81–114.

19. Hornaday, *Our Vanishing Wild Life,* 212.

20. Edward Forbush, *Useful Birds and Their Protection* (Boston: Wright and Potter, 1908), 51–72.

21. Hornaday, *Our Vanishing Wild Life,* 221–22.

22. For American attitudes toward vermin, see especially Dunlap, *Saving America's Wildlife,* 34–61. His book focuses on the Biological Survey's attitude toward predators in general, not just birds.

23. George A. Lawyer, *Federal Protection of Migratory Birds,* Yearbook of the Department of Agriculture, No. 785 (Washington, DC: U.S. Government Printing Office, 1919), 4–5.

24. Doughty, *Feather Fashions,* 106–7.

25. Connery, *Governmental Problems,* 54–63 (quote on 54). For a full analysis of *Geer v. Connecticut,* see Dale D. Goble and Eric T. Freyfogle, *Wildlife Law: Cases and Materials* (New York: Foundation Press, 2002), 387–96, 427–34.

26. T. S. Palmer, W. F. Bancroft, and Frank L. Earnshaw, "Game Laws of 1913: A Summary of the Provisions Relating to Seasons, Export, Sale, Limits, and Licenses," *Bulletin of the U.S. Department of Agriculture,* no. 22 (September 16, 1913): 10.

27. Ibid., 38–50.

28. Ibid., 10.

29. Ibid., 11–18; Lawyer, *Federal Protection of Migratory Birds,* 4–5; and Doughty, *Feather Fashions,* 106–7.

30. See Connery, *Governmental Problems,* 31–80, for a full discussion of federal-state issues.

31. Phillips, *Migratory Bird Protection in North America*, 7–8. For a complete analysis of the Lacey Act's influence on twentieth-century wildlife law, see Goble and Freyfogle, *Wildlife Law*, 831–50.

32. United States Department of Agriculture, *Service and Regulatory Announcements (August 21, 1916)*, Bureau of Biological Survey S.R.A.-B.S. 11 (Washington, DC: U.S. Government Printing Office, 1916) (italics added).

33. Doughty, *Feather Fashions*, 125–34 (quote on 131).

34. Ossa, *They Saved Our Birds*, 57–58.

35. Phillips, *Migratory Bird Protection in North America*, 10–13.

36. Senate Committee on Agriculture and Forestry, *Agriculture Appropriation Bill, 1917 (Protection of Migratory Birds): Hearings on H. R. 12717*, 64th Cong., 1st sess., 1916, 41.

37. Sherman Strong Hayden, *The International Protection of Wild Life* (New York: Columbia University Press, 1942), 73n23. Hayden mistakenly calls this case *State v. McCullagh.*

38. Senate Committee, *Agriculture Appropriation Bill*, 4–13 (quote on 8).

39. Alexander M. Bickel and Benno C. Schmidt, eds., *The Judiciary and Responsible Government 1910–21*, vol. 9 of *History of the Supreme Court of the United States* (New York: Macmillan, 1984), 476–82; and Goble and Freyfogle, *Wildlife Law*, 517–22.

40. Historical scholarship on the 1916 U.S.-Canadian bird treaty is sparse. By far the best overall survey is Kurkpatrick Dorsey, *The Dawn of Conservation Diplomacy: U.S.-Canadian Wildlife Protection Treaties in the Progressive Era* (Seattle: University of Washington Press, 1998), 164–214. See also Trefethen, *American Crusade for Wildlife*, 143–56; and Janet Foster, *Working for Wildlife: The Beginning of Preservation in Canada* (Toronto, Canada: University of Toronto Press, 1998), 120–48.

41. Phillips, *Migratory Bird Protection in North America*, 13.

42. Cees Flinterman, Barbara Kwiatkowska, and Johan G. Lammers, eds., *Transboundary Air Pollution: International Legal Aspects of the Co-operation of States* (Dordrecht, the Netherlands: Martinus Nijhoff, 1986), 13.

43. McLean's resolution in Wilson's letter to Bryan of April 19, 1913, box 7230, decimal file 800.6232/140 (1910–1929), RG 59, NARA.

44. Wilson's letter to Bryan on April 19, 1913, box 7230, decimal file 800.6232/140 (1910–1929), RG 59, NARA.

45. Quote is from letter from Acting Secretary of Agriculture (Beverly Thomas Galloway) to Secretary of State on May 2, 1913, box 7230, decimal file 800.6232/140 (1910–1929), RG 59, NARA. The rest of the material is from the letter from Ambassador Spring-Rice to Acting Secretary of State J. B. Moore on October 20, 1913, box 5528, decimal file 562.23A8/85 (1910–1929), RG 59,

NARA. The United States became interested in the British-backed initiative after the Underwood Tariff of 1913 was passed.

46. See Dorsey, *Dawn of Conservation Diplomacy,* 193–216.

47. See Alexander J. Burnett, *A Passion for Wildlife: The History of the Canadian Wildlife Service* (Vancouver, Canada: UBC Press, 2003); Foster, *Working for Wildlife*; and Tina Loo, *States of Nature: Conserving Canada's Wildlife in the Twentieth Century* (Vancouver, Canada: University of British Columbia Press, 2006). Both Burnett and Foster emphasize the top-down nature of Canadian conservation, whereas Loo highlights the activities of Jack Miner and many others who made individual contributions to nature protection.

48. Phillips, *Migratory Bird Protection in North America,* 16–17; and Foster, *Working for Wildlife,* 3–8 and 124–32.

49. See "Treaty Would Kill Game Shooting in This Province" and "Contemplated Treaty Strongly Criticized," both in Vancouver's daily newspaper, *Province,* August 29, 1916.

50. See Phillips, *Migratory Bird Protection in North America,* 17–20; C. Gordon Hewitt, "Conservation of Birds and Mammals in Canada," in *Conservation of Fish, Birds and Game: Proceedings at a Meeting of the Committee, November 1 and 2, 1915* (Toronto, Canada: Methodist Book and Publishing House, 1916), 141–43; and Foster, *Working for Wildlife,* 134–36.

51. David Franklin Houston, "Memorandum of Suggested Changes in the Convention between the United States and Great Britain for the Protection of Migratory Birds in the United States and Canada (10 March 1916)," box 7230, decimal file 800.6232/140 (1910–1929), RG 59, NARA.

52. Ibid.

53. Ibid. (italics added to indicate new text).

54. Ibid.

55. Ibid.

56. Ibid.

57. Ibid.; Foster, *Working for Wildlife,* 141–42.

58. Houston, "Memorandum of Suggested Changes."

59. Ibid. (italics added to indicate new text).

60. Dan Gottesman, "Native Hunting and the Migratory Birds Convention Act: Historical, Political and Ideological Perspectives," *Journal of Canadian Studies* 18, no. 3 (1983): 71–80; and E. Brian Titley, *A Narrow Vision: Duncan Campbell Scott and the Administration of Indian Affairs in Canada* (Vancouver, Canada: University of British Columbia Press, 1986), esp. 53–55.

61. *Protocol Amending the 1916 Convention for the Protection of Migratory Birds.* 104th Cong., 2d sess., 1996, S. Treaty Doc. 28. At that time, negotiators

also updated and simplified the text of the 1916 Convention, without altering its scope or intent.

62. Houston, "Memorandum of Suggested Changes."

63. Letter from British Embassy in Washington, DC, to Secretary of State Robert Lansing on February 12, 1916, box 7230, decimal file 800.6232/140 (1910–1929), RG 59, NARA.

64. Hayden, *International Protection of Wild Life*, 78–79; Dorsey, *Dawn of Conservation Diplomacy*, 215–31; Goble and Freyfogle, *Wildlife Law*, 852–57.

65. U.S. Department of Agriculture, *The Migratory Bird Treaty: Decision of the Supreme Court of the United States Sustaining the Constitutionality of the Migratory Bird Treaty and Act of Congress to Carry It into Effect*, Department Circular 102 (Washington, DC: U.S. Department of Agriculture, 1920), 3–4. See also Goble and Freyfogle, *Wildlife Law*, 522–26.

66. U.S. Department of Agriculture, *The Migratory Bird Treaty Act: "United States vs. Joseph H. Lumpkin,"* department circular 202 (Washington, DC: U.S. Department of Agriculture, 1922), 3.

67. Foster, *Working for Wildlife*, 147; Dorsey, *Dawn of Conservation Diplomacy*, 233.

68. Foster, *Working for Wildlife*, 155–60, 202–9 (quote on 204).

69. Titley, *Narrow Vision*, 55.

70. Lawyer, *Federal Protection of Migratory Birds*, 7.

71. Canadian Wildlife Service, *List of Birds Protected in Canada under the Migratory Birds Convention Act* (Ottawa: Ministry of Fisheries and the Environment, 1978), 7–32. The most up-to-date list can be found at http://migratorybirds.fws.gov/intrnltr/mbta/mbtintro.html.

72. William T. Hornaday, *Thirty Years War for Wild Life: Gains and Losses in the Thankless Task* (Stamford, CT: Permanent Wild Life Protection Fund, 1931), 116. Hornaday published a map of duck- and goose-hunting seasons for 1929.

73. Lawyer, *Federal Protection of Migratory Birds*, 7.

74. Robert P. Allen, "The Wild-Life Sanctuary Movement in the United States," *Bird-Lore* 36 (1934): 82; Trefethen, *American Crusade for Wildlife*, 122–24; and United States. Department of Agriculture, *National Reservations for the Protection of Wild Life*, department circular 87 (Washington, DC: U.S. Department of Agriculture, 1912). The Department of Agriculture circular provides a list of these refuges (which is also available in Hornaday, *Our Vanishing Wild Life*, 345).

75. Ira N. Gabrielson, *Wildlife Refuges* (New York: Macmillan, 1943), 7.

76. T. Gilbert Pearson, "What Is the Audubon Society?" *National Association of Audubon Societies*, circular no. 16 (1930).

77. House Committee on Agriculture, *Migratory Bird Refuges and Public Shooting Grounds: Hearings,*70th Cong., 2nd sess., 1922, Series T, 36–37.

78. Trefethen, *American Crusade for Wildlife*, 185–89.

79. Paul Redington, *Report of the Chief of the Bureau of Biological Survey, U.S. Department of Agriculture* (Washington, DC: U.S. Government Printing Office, 1931), 37. See also Senate Committee on Agriculture and Forestry, *Bear River Migratory Bird Refuge: Hearings on S. 703, a Bill to Establish the Bear River Migratory Bird Refuge and S. 1272, a Bill Authorizing the Establishment of a Migratory Bird Refuge at Bear River Bay, Great Salt Lake, Utah,* 70th Cong., 1st sess., 1928.

80. Trefethen, *American Crusade for Wildlife*, 188–89. The appropriation was later reduced in the wake of the Great Depression.

81. Hayden, *International Protection of Wild Life*, 80–85; Connery, *Governmental Problems*, 102–8.

82. "Report of the Migratory Bird Conservation Commission for the Fiscal Year 1954," box 460, Records of the Office of the Secretary of the Interior, Central Classified Files, 1954–58, RG 48, NARA.

83. Lynn A. Greenwalt, "The National Wildlife Refuge System," in *Wildlife and America: Contributions to an Understanding of American Wildlife and Its Conservation*, ed. Howard P. Brokaw, 399–412 (Washington, DC: U.S. Government Printing Office, 1978). Greenwalt offers the best overall history, but it only goes up to 1978. For the most up-to-date list of refuge areas, see http://refuges100.fws.gov/facts/chronology.htm.

84. Trefethen, *American Crusade for Wildlife*, 217–29, offers a good sketch of the Biological Survey's activities, especially under "Ding" Darling in the 1930s.

85. House Committee on Agriculture, *Amend the Migratory Bird Treaty Act with Respect to Bag Limits: Hearings on H. R. 5278 by Mr. Haugen*, 71st Cong., 2nd sess., 1930, Series M, 3.

86. Phillips, *Migratory Bird Protection in North America*, 27.

87. Hewitt, "Conservation of Birds and Mammals," 300–309.

88. Foster, *Working for Wildlife*, 185–92.

89. Hewitt, "Conservation of Birds and Mammals," 146–48; Jack Miner, *Jack Miner and the Birds, and Some Things I Know about Nature* (Chicago: Reilly and Lee, 1923).

90. *The Migratory Birds Convention Act and Federal Regulations for the Protection of Migratory Birds (Revised Statutes of Canada)* (Ottawa: F. A. Acland, 1933). This pamphlet lists the 1919 revisions and the reserves.

91. Phillips, *Migratory Bird Protection in North America*, 27.

92. Albert M. Day, *North American Waterfowl* (New York: Stackpole and Heck, 1949), 255.

93. Figures are from Environment Canada's website, http://www.mb.ec.gc.ca/nature/migratorybirds/sanctuaries/dc01s00.en.html.

94. House of Commons (Canada), *House of Commons Debates: Official Report—Unrevised Edition, Friday, April 20, 1934* (Ottawa: J. O. Patenaude, 1934), 2594.

95. Lane Simonian, *Defending the Land of the Jaguar: A History of Conservation in Mexico* (Austin: University of Texas Press, 1995), 79–96 (quote on 79); and Fernando Vargas Márquez, *Los parques nacionales de México y reservas equivalentes: Pasado, presente y futuro* (Mexico City: Instituto de Investigaciónes Económicas–UNAM, 1984), 43–49.

96. Letter from Wallace to President Wilson of April 9, 1913, box 7230, decimal file 800.6232/140 (1910–1929), RG 59, NARA. For a more detailed analysis of the discussions within the U.S. government about the prospects of a treaty with Mexico, see Dorsey, *Dawn of Conservation Diplomacy*, 194–98.

97. Letter from John H. Wallace to Acting Secretary of State Frank L. Polk of February 14, 1919, box 7230, decimal file 800.6232 (1910–1929), RG 59, NARA.

98. Memorandum from L. J. C. to Phillips of November 14, 1918, box 7230, decimal file 800.6232 (1910–1929), RG 59, NARA.

99. Letter from Department of Agriculture to Department of State, October 11, 1924, box 7230, decimal file 800.6232/140 (1910–1929), RG 59, NARA.

100. Letter from Acting Secretary of Agriculture R. W. Dunlap to Department of State, November 13, 1925, box 7230, decimal file 800.6232/140 (1910–1929), RG 59, NARA.

101. Letter from Department of Agriculture to Department of State, November 10, 1934, box 4625, decimal file 800.6232/140 (1930–1939), RG 59, NARA.

102. Letter from Manuel C. Tellez to the Secretary of the Department of State, August 2, 1929, box 3, Records of the U.S. Fish and Wildlife Service, General Records, Records Concerning Relations with Mexico, 1925–1937, RG 22, NARA.

103. The text of this agreement is in Bernd Rüster and Bruno Simma, eds., *International Protection of the Environment: Treaties and Related Documents* (Dobbs Ferry, NY: Oceana Publications, 1975), 4:1723–24.

104. Ibid., 1723.

105. Ibid.

106. Ibid., 1724.

107. Ibid.

108. Text in ibid., 5:2217–18.

109. *Protocol with Mexico Amending Convention for Protection of Migratory Birds and Game Mammals*, 105th Cong., 1st sess., 1997, S. Treaty Doc. 26.

110. Letter from State Department to Mrs. Charles N. (Rosalie) Edge, Chair of the Emergency Conservation Committee of New York, May 4, 1936, SEN 74B-B15, RG 46, NARA.

111. Simonian, *Defending the Land of the Jaguar,* 128–29.

112. Luis Macias, "Wildlife Problems in Mexico," in *Transactions of the Fourteenth North American Wildlife Conference* (Washington, DC: Wildlife Management Institute, 1949), 11.

113. George B. Saunders, "Waterfowl Wintering Grounds of Mexico," in *Transactions of the Seventeenth North American Wildlife Conference* (Washington, DC: Wildlife Management Institute, 1952), 90, 98.

114. Ibid., 92.

115. The reports are in box 4625, decimal file 800.6232/156–984 (1930–39), RG 59, NARA.

116. American Committee for International Wild Life Protection, *Brief History and Text of the Convention on Nature Protection and Wild Life Preservation in the Western Hemisphere* (New York: American Committee for International Wild Life Protection, 1946); and Keri Lewis, *Negotiating Nature: The Convention on Nature Protection and Wildlife Preservation in the Western Hemisphere, 1930–1950* (PhD diss., University of New Hampshire, 2007).

117. L. S. Rowe and Pedro de Alba, eds., *Preliminary Draft Convention on Nature Protection and Wild Life Preservation in the Western Hemisphere* (Washington, DC: Pan American Union, 1940). Final text is in Rüster and Simma, *International Protection of the Environment,* 5:1729–32 (italics added). Regarding migratory birds, the Preliminary Draft and the Final Text were identical.

118. Rüster and Simma, *International Protection of the Environment,* 4:1739–60.

119. Hornaday, *Thirty Years War for Wild Life,* 1. Hornaday's emphasis.

120. On the history of the poultry industry, see especially William Boyd, "Making Meat: Science, Technology, and American Poultry Production," *Technology and Culture* 42, no. 4 (October 2001): 631–64. The industry began in the 1890s, but it really boomed in the 1920s, as the supply of wild birds dried up.

121. Hornaday, *Thirty Years War for Wild Life,* 36.

122. Frederick Lincoln, *The Migration of American Birds* (New York: Doubleday, 1939), 31–91; Jean Dorst, *The Migrations of Birds* (Boston: Houghton Mifflin, 1962), 97–139.

123. T. E. Dahl, *Wetlands Losses in the United States 1780's to 1980's* (Washington, DC: U.S. Department of the Interior, Fish and Wildlife Services, 1990), 5–6.

124. W. A. Glooschenko, C. Tarnocai, S. Zoltai, and V. Gloosehenko "Wetlands of Canada and Greenland," in *Wetlands of the World: Inventory, Ecology and Management,* ed. Dennis Whigham, Dagmar Dykyjová, and Slavomil Hejný, 496–98 (Dordrecht, the Netherlands: Kluwer, 1993).

125. Ingrid Olmsted, "Wetlands of Mexico," in Whigham, Dykyjová, and Hejný, *Wetlands of the World,* 670–73.

Chapter 3: The Antarctic Whale Massacre

1. On the history of world whaling, see especially J. N. Tønnessen and A. O. Johnsen, *The History of Modern Whaling* (Berkeley: University of California Press, 1982); Richard Ellis, *Men and Whales* (New York: Alfred A. Knopf, 1991); and Daniel Francis, *A History of World Whaling* (Harmondsworth, UK: Penguin, 1990). See also J. T. Jenkins, *A History of the Whale Fisheries from the Basque Fisheries of the Tenth Century to the Hunting of the Finner Whale at the Present Date* (Port Washington, NY: Kennikat Press, 1921); R. B. Robertson, *Of Whales and Men* (New York: Alfred A. Knopf, 1961); George L. Small, *The Blue Whale* (New York: Columbia University Press, 1971); K. Radway Allen, *Conservation and Management of Whales* (Seattle: University of Washington Press, 1980); Patricia Birnie, comp. and ed., *International Regulation of Whaling: From Conservation of Whaling to Conservation of Whales and Regulation of Whale-Watching* (New York: Oceana Publications, 1985); Jeremy Cherfas, *The Hunting of the Whale* (London: Penguin, 1989); Peter J. Stoett, *The International Politics of Whaling* (Vancouver, Canada: University of British Columbia Press, 1997); and Granville Allen Mawer, *Ahab's Trade: The Saga of South Seas Whaling* (New York: St. Martin's Press, 1999).

2. Tønnessen and Johnsen, *History of Modern Whaling*, 28–36; Ellis, *Men and Whales*, 255–65.

3. Tønnessen and Johnsen, *History of Modern Whaling*, 28–36; Ellis, *Men and Whales*, 255–65.

4. Tønnessen and Johnsen, *History of Modern Whaling*, 386.

5. Ibid., 25, 39, 69.

6. Ellis, *Men and Whales*, 47.

7. Tønnessen and Johnsen, *History of Modern Whaling*, 68.

8. Small, *Blue Whale*, 11–12; and Tønnessen and Johnsen, *History of Modern Whaling*, 313.

9. For good introductions to cetacean biology, see Peter G. H. Evans, *The Natural History of Whales and Dolphins* (New York: Facts on File Publications, 1987), 1–19; Leonard Harrison Matthews, *The Natural History of the Whale* (New York: Columbia University Press, 1978), 23–47.

10. Evans, *Natural History of Whales and Dolphins*, 1–19; Matthews, *Natural History of the Whale*, 23–47.

11. See the introductory remarks of Janet Mann, Richard C. Connor, Peter L. Tyack, and Hal Whitehead, eds., *Cetacean Societies: Field Studies of Dolphins and Whales* (Chicago: University of Chicago Press, 2000), 11–17; Matthews, *Natural History of the Whale*, 48–75; and Richard C. Connor and Dawn Micklethwaite Peterson, *The Lives of Whales and Dolphins* (New York: Henry Holt, 1994), 11–17.

12. George A. Knox, *The Biology of the Southern Ocean* (Cambridge: Cambridge University Press, 1994), 167–68; Hal Whitehead and Janet Mann, "Female Reproductive Strategies of Cetaceans: Life Histories and Calf Care," in Mann, Connor, Tyack, and Whitehead, *Cetacean Societies,* 219–46. Knox's account is succinct, but Whitehead's contains more detail and includes dolphins and porpoises.

13. Knox, *Biology of the Southern Ocean,* 167–68; Whitehead and Mann, "Female Reproductive Strategies of Cetaceans."

14. Leonard Harrison Matthews, *The Whale* (New York: Simon and Schuster, 1968), 160–86; and Matthews, *Natural History of the Whale,* 1–22.

15. Phillip J. Clapham, "The Humpback Whale: Seasonal Feeding and Breeding in a Baleen Whale," in Mann, Connor, Tyack, and Whitehead, *Cetacean Societies,* 173–96.

16. Hal Whitehead and Linda Weilgart, "The Sperm Whale: Social Females and Roving Males," in Mann, Connor, Tyack, and Whitehead, *Cetacean Societies,* 154–72; Lance E. Davis, Robert E. Gallman, and Teresa D. Hutchins, "The Decline of U.S. Whaling: Was the Stock of Whales Running Out?" *Business History Review* 62, no. 4 (1988): 569–95.

17. Ellis, *Men and Whales,* 392.

18. Åge Jonsgård, *Biology of the North Atlantic Fin Whale Balaeoptera Physalus (L): Taxonomy, Distribution, Migration and Food* (Oslo: Universitetsforlaget, 1966).

19. Amy Samuels and Peter Tyack, "Flukeprints: A History of Studying Cetacean Societies," in Mann, Connor, Tyack, and Whitehead, *Cetacean Societies,* 10–12.

20. Tønnessen and Johnsen, *History of Modern Whaling,* 367.

21. Matthews, *Whale,* 220–22; Karl Brandt, *Whaling and Whale Oil during and after World War II,* War-Peace Pamphlets (Stanford, CA: Food Research Institute, 1948), 1–3; and Gordon Jackson, *The British Whaling Trade* (London: Adam and Charles Black, 1978), 178–86.

22. Matthews, *Whale,* 236.

23. Ibid., 222–23.

24. Ellis, *Men and Whales,* viii, 144–45.

25. Francis, *History of World Whaling,* 17–27; Ellis, *Men and Whales,* 42–47.

26. Jenkins, *History of the Whale Fisheries,* 59–118; Ellis, *Men and Whales,* 48–71.

27. Jenkins, *History of the Whale Fisheries,* 119–76; Ellis, *Men and Whales,* 57–71, 224.

28. Ellis, *Men and Whales,* 80–89.

29. Lance E. Davis, Robert E. Gallman, and Karin Gleiter, *In Pursuit of Leviathan: Technology, Institutions, Productivity, and Profits in American Whaling,*

1816–1906 (Chicago: University of Chicago Press, 1997); Alexander Starbuck, *The History of the American Whale Fishery* (Secaucus, NJ: Castle Books, 1989 [1877]).

30. On Roys, see Frederick P. Schmitt, Cornelis de Jong, and Frank H. Winter, *Thomas Welcome Roys: America's Pioneer of Modern Whaling* (Charlottesville: University Press of Virginia, 1980). See also Davis, Gallman, and Gleiter, *In Pursuit of Leviathan*, 34–46, 260–96; Ellis, *Men and Whales*, 141–46, 158–62, 256–57; and Starbuck, *History of the American Whale Fishery*.

31. Cited by Frederick P. Schmitt, *Mark Well the Whale! Long Island Ships to Distant Seas* (Port Washington, NY: Kennikat Press, 1971), 111.

32. Tønnessen and Johnsen, *History of Modern Whaling*, 16.

33. Jenkins, *History of the Whale Fisheries*, 238.

34. Jackson, *British Whaling Trade*, 3–154.

35. Tønnessen and Johnsen, *History of Modern Whaling*, 33.

36. Committee for Whaling Statistics, *International Whaling Statistics II* (Oslo: Committee for Whaling Statistics, 1931), 6–9. This record offers the most complete set of data available, but it is still incomplete.

37. Tønnessen and Johnsen, *History of Modern Whaling*, 68, and more fully on 75–110.

38. Ibid., 111–46 (statistics on 733–35).

39. Francis, *History of World Whaling*, 185–91; Ellis, *Men and Whales*, 351–52.

40. Tønnessen and Johnsen, *History of Modern Whaling*, 157–78.

41. Ibid., 226.

42. Ibid., 176, 313.

43. Small, *Blue Whale*, 11–12.

44. An Australian correspondent, A. J. Villiers, chronicled this adventure. See A. J. Villiers, *Whaling in the Frozen South: Being the Story of the 1923–24 Norwegian Whaling to the Antarctic* (New York: Bobbs-Merrill, 1931 [1925]).

45. *Pelagic whaling* is the term most commonly used to describe Antarctic hunting, but it is something of a misnomer. Over the centuries, various hunters of different nationalities had killed whales on the high seas, that is, pelagically. This marked the first time that whalers penetrated into the ice blanket of the polar regions and then returned to their home ports without first unloading their oil at a shore station or shore-based floating factory. *Ice whaling* is a more accurate term.

46. Tønnessen and Johnsen, *History of Modern Whaling*, 313; J. L. McHugh, "Rise and Fall of World Whaling: The Tragedy of the Commons Illustrated," *Journal of International Affairs* 31, no. 1 (1977): 23–26. As McHugh explains,. the 1931 catch translated into 34,668 BWU and the 1938 catch into 29,786 BWU.

47. Tønnessen and Johnsen, *History of Modern Whaling*, 61–67, 81–85, 304–5.

48. Ibid., 141–42.

49. Ibid., 90, 104, 109.

50. Ibid., 194, 208–19, 319; A. Gruvel, "La chasse aux cétacés dans le monde: Son avenir dans les colonies françaises," *Revue scientifique: Revue rose* 63, no. 3 (February 1925): 65–71.

51. Tønnessen and Johnsen, *History of Modern Whaling,* 180–81, 185–86, 340.

52. Birnie, *International Regulation of Whaling,* 109–13 (quote on 112).

53. Tønnessen and Johnsen, *History of Modern Whaling,* 364–66; Small, *Blue Whale,* 145–52.

54. M. Braadland (Norway), "Report and Draft Resolution Presented by the Second Commission to the Assembly," League of Nations: Opening of a Convention for the Regulation of Whaling, decimal file 562.8F 1/25 (1930–39), RG 59, NARA.

55. Bernd Rüster and Bruno Simma, eds., *International Protection of the Environment: Treaties and Related Documents* (Dobbs Ferry, NY: Oceana Publications, 1975), 7:3469.

56. Tønnessen and Johnsen, *History of Modern Whaling,* 403–4.

57. Rüster and Simma, *International Protection of the Environment,* 7:3469–73.

58. Ibid., 3469–70. Rüster and Simma include the final text, but the pertinent information on the debates and amendments come from "Summary of the Replies of Governments with Regard to the Preliminary Draft Convention for the Regulation of Whaling (1931)," PRO FO 371/15658/186629.

59. "Summary of the Replies of Governments."

60. Department of State memo, "Convention for the Regulation of Whaling" (February 23, 1932), decimal file 562.8F 1/47 (1930–39), RG 59, NARA.

61. "Note by the League of Nations Secretary-General, Draft Convention for the Regulation of Whaling" (September 12, 1931), PRO FO 371/15659.

62. See "Telegram from the Secretary of State for Dominion Affairs to the Governor General of New Zealand" (July 30, 1931), PRO FO 371/15659.

63. Karl Brandt, *Whale Oil: An Economic Analysis* (Stanford, CA: Food Research Institute, 1940), 61; and Tønnessen and Johnsen, *History of Modern Whaling,* 385.

64. Tønnessen and Johnsen, *History of Modern Whaling,* 386–94.

65. Birnie, *International Regulation of Whaling,* 120–24.

66. Ibid., 122–24; McHugh, "Rise and Fall of World Whaling," 26.

67. Tønnessen and Johnsen, *History of Modern Whaling,* 370. See also "Note of an Interview with Mr. F. D'Arcy Cooper, Chairman of Uni-Lever Ltd. (19 January 1931)" PRO FO 371/15671 (1931), and "Preliminary Whaling Conference, Oslo, May 1938," PRO FO 371/22273/187067 (1938).

68. "Preliminary Whaling Conference."

69. Birnie, *International Regulation of Whaling*, 122–24.

70. Ellis, *Men and Whales*, 367–70, 386–87; Sherman Strong Hayden, *The International Protection of Wild Life* (New York: Columbia University Press, 1942), 166–67; and Tønnessen and Johnsen, *History of Modern Whaling*, 329. See also the *Japanese Advertiser* (August 12, 1973, and December 31, 1938), in which Japan proclaimed its intention of breaking the Norwegian-British "monopoly" before it would sign any whaling treaties.

71. For the original British draft, see "International Conference on Whaling, Draft Agreement for the Regulation of the Industry of Whaling," PRO FO 371/21078 (1937). It does not differ significantly from the final draft. Until the 1946 treaty was signed, this agreement was more commonly known as the Principal Agreement.

72. Rüster and Simma, *International Protection of the Environment*, 3475–76.

73. Tønnessen and Johnsen, *History of Modern Whaling*, 453.

74. Rüster and Simma, *International Protection of the Environment*, 3476–77.

75. Ibid., 3477–78.

76. Hayden, *International Protection of Wild Life*, 160–61.

77. Tønnessen and Johnsen, *History of Modern Whaling*, 462.

78. "International Whaling Conference (I.W.C. Paper No. 3)," PRO FO 372/4380, Treaty Formalities (1945). See also "Report on the International Whaling Conference Held in London November 20 through November 26, 1945," box 7, Records Relating to International Whaling Conferences (1936–49), RG 43, NARA.

79. Tønnessen and Johnsen, *History of Modern Whaling*, 457.

80. *Financial News*, December 1, 1937.

81. Tønnessen and Johnsen, *History of Modern Whaling*, 489–91. See also "Report on the International Whaling Conference."

82. For good overviews of the 1946 Washington Convention, see Simon Lyster, *International Wildlife Law: An Analysis of International Treaties Concerned with the Conservation of Wildlife* (Cambridge: Grotius Publications, 1985), 17–38; and Gregory Rose and Saundra Crane, "The Evolution of International Whaling Law," in *Greening International Law*, ed. Philippe Sands, 159–81 (New York: New Press, 1994).

83. Rüster and Simma, *International Protection of the Environment*, 3489.

84. "United States Proposals for a Whaling Convention," box 8, Records Relating to International Whaling Conferences (1936–49), RG 43, NARA.

85. Rüster and Simma, *International Protection of the Environment*, 3498–3511.

86. Steinar Andresen, "Science and Politics in the International Management of Whales," *Marine Policy* 13, no. 2 (April 1989): 103.

87. Rüster and Simma, *International Protection of the Environment*, 3501–3.

Compare with "United States Proposals for a Whaling Convention," box 8, Records Relating to International Whaling Conferences (1936–49), RG 43, NARA.

88. Rüster and Simma, *International Protection of the Environment*, 3501 and 3504.

89. "Minutes of the Fourteenth Session, Saturday, November 30, 1046," box 8, Records Relating to International Whaling Conferences, 1936–1949, RG 43, NARA.

90. G. H. Elliot, "The Failure of the IWC, 1946–1966," *Marine Policy* 3, no. 2 (April 1979): 152.

91. Lyster, *International Wildlife Law*, 27.

92. Elliot, "The Failure of the IWC," 152.

93. Ibid., 152–53.

94. Ellis, *Men and Whales*, 431–33.

95. Ian G. Barbour, ed., *Western Man and Environmental Ethics: Attitudes toward Nature and Technology* (Reading, MA: Addison-Wesley, 1973), 104.

96. Cited by Ellis, *Men and Whales*, 390.

97. Cited by David Day, *The Whale War* (London: Routledge and Kegan Paul, 1987), 136.

98. Jackson, *British Whaling Trade*, 239.

99. Cited by Ellis, *Men and Whales*, 388–89.

100. Hermann Melville, *Moby Dick; or, The Whale* (Indianapolis, IN: Bobbs-Merrill, 1964 [1851]), 586.

Conclusion

1. Aldo Leopold, *Round River: From the Journals of Aldo Leopold*, ed. Luna B. Leopold (New York: Oxford University Press, 1953), 145.

2. I. S. C. Parker and Mohamed Amin, *Ivory Crisis* (London: Chatto and Windus, 1983), 20–23.

3. These data are from "Ramsar List of Wetlands of International Importance," Secretariat of the Convention on Wetlands (Ramsar, Iran, 1971), Rue Mauverney, CH 1196, Gland, Switzerland. Also available at http://www.ramsar .org/key_sitelist.htm.

4. Simon Lyster, *International Wildlife Law: An Analysis of International Treaties Concerned with the Conservation of Wildlife* (Cambridge: Grotius Publications, 1985), 19.

Bibliography

Unpublished Archival Records

National Archives and Records Administration (NARA), College Park, MD. Record Groups (RG) 22, 43, 46, 48, 59.
Public Records Office (PRO), Kew Gardens, United Kingdom. Records of the Foreign Office (FO). Africa.

Published Material

Adams, Jonathan S., and Thomas O. McShane. *The Myth of Wild Africa: Conservation without Illusion.* New York: W. W. Norton, 1992.

Adams, William M. *Against Extinction: The Story of Conservation.* London: Earthscan, 2004.

Allen, K. Radway. *Conservation and Management of Whales.* Seattle: University of Washington Press, 1980.

Allen, Robert P. "The Wild-Life Sanctuary Movement in the United States." *Bird-Lore* 36 (1934): 80–84.

Allsen, Thomas T. *The Royal Hunt in Eurasian History.* Philadelphia: University of Pennsylvania Press, 2006.

Alpers, Edward A. *Ivory and Slaves: Changing Patterns of International Trade in East Central Africa to the Later Nineteenth Century.* Berkeley: University of California Press, 1975.

American Committee for International Wild Life Protection. *African Game Protection: An Outline of the Existing Game Reserves and National Parks of Africa.* Cambridge, MA: American Committee for International Wild Life Protection, 1933.

———. *Brief History and Text of the Convention on Nature Protection and Wild Life Preservation in the Western Hemisphere.* New York: American Committee for International Wild Life Protection, 1946.

———. *The London Convention for the Protection of African Fauna and Flora with Map and Notes on Existing Parks and Reserves*. Cambridge, MA: American Committee for International Wild Life Protection, 1935.

Anderson, David, and Richard Grove, eds. *Conservation in Africa: Peoples, Policies and Practice*. Cambridge: Cambridge University Press, 1987.

Andresen, Steinar. "Science and Politics in the International Management of Whales." *Marine Policy* 13, no. 2 (April 1989): 99–117.

Askins, Robert A. *Restoring North America's Birds: Lessons from Landscape Ecology*. New Haven, CT: Yale University Press, 2000.

Atwood, Wallace W. *The Protection of Nature in the Americas*. Instituto Panamericano de Geografía e Historia. Mexico City: Antigua Imprenta de E. Murguía, 1940.

Barbour, Ian G., ed. *Western Man and Environmental Ethics: Attitudes toward Nature and Technology*. Reading, MA: Addison-Wesley, 1973.

Barcley, E. N. *Big Game Shooting Records*. London: Thames and Hudson, 1956.

Barrow, Mark V. *A Passion for Birds: American Ornithology after Audubon*. Princeton, NJ: Princeton University Press, 1998.

———. "Science, Sentiment, and the Specter of Extinction: Reconsidering Birds of Prey during America's Interwar Years." *Environmental History* 7, no. 1 (January 2002): 69–98.

Beachey, R. W. *The Slave Trade of Eastern Africa*. New York: Barnes and Noble, 1976.

Bean, M. J. *The Evolution of National Wildlife Law*. Washington, DC: Council on Environmental Quality, 1977.

Beard, Peter. *The End of the Game*. New York: Viking, 1965.

Beinart, William. *The Rise of Conservation in South Africa: Settlers, Livestock, and the Environment 1770–1950*. Oxford: Oxford University Press, 2003.

Beinart, William, and Peter Coates. *Environment and History: The Taming of Nature in the USA and South Africa*. London: Routledge, 1995.

Belanger, Dian Olson. *Managing American Wildlife: A History of the International Association of Fish and Wildlife Agencies*. Amherst: University of Massachusetts Press, 1988.

Bennett, A. G. *Whaling in the Antarctic*. Edinburgh: Wm. Blackwood and Sons, 1931.

Bickel, Alexander M., and Benno C. Schmidt, eds. *The Judiciary and Responsible Government 1910–21*. Vol. 9 of *History of the Supreme Court of the United States*. New York: Macmillan, 1984.

Birnie, Patricia, comp. and ed. *International Regulation of Whaling: From Conservation of Whaling to Conservation of Whales and Regulation of Whale-Watching*. New York: Oceana Publications, 1985.

Boahen, A. Adu. *Africa under Colonial Domination 1880–1935.* Berkeley: University of California Press, 1985.

Boardman, Robert. *International Organization and the Conservation of Nature.* Bloomington: Indiana University Press, 1981.

Bonner, Raymond. *At the Hand of Man: Peril and Hope for Africa's Wildlife.* New York: Alfred A. Knopf, 1993.

Boyd, William. "Making Meat: Science, Technology, and American Poultry Production." *Technology and Culture* 42, no. 4 (October 2001): 631–64.

Brandt, Karl. *Whale Oil: An Economic Analysis.* Stanford, CA: Food Research Institute, 1940.

———. *Whaling and Whale Oil during and after World War II.* War-Peace Pamphlets. Stanford, CA: Food Research Institute, 1948.

Brooks, P. M., and I. A. W. Macdonald. "The Hluhluwe-Umfolozi Reserve: An Ecological Case History." In *Management of Large Mammals in African Conservation Areas: Proceedings of a Symposium Held in Pretoria, South Africa, 29–30 April 1982,* edited by R. Norman Owen-Smith, 51–77. Pretoria, South Africa: HAUM, 1983.

Bryden, H. A. "The Extermination of Game in South Africa." *Fortnightly Review* 62 (1894): 538–51.

———, ed. *Great and Small Game of Africa: An Account of the Distribution, Habits, and Natural History of the Sporting Mammals, with Personal Hunting Experiences.* London: Rowland Ward, 1899.

Burnett, Alexander J. *A Passion for Wildlife: The History of the Canadian Wildlife Service.* Vancouver, Canada: UBC Press, 2003.

Burton, Robert. *The Life and Death of Whales.* Totowa, NJ: Rowman and Allanheld, 1983.

Canada. Parliament. House of Commons. *House of Commons Debates. Official Report—Unrevised Edition, Friday, April 20, 1934.* Ottawa: J. O. Patenaude, 1934.

Canadian Wildlife Service. *List of Birds Protected in Canada under the Migratory Birds Convention Act.* Ottawa: Ministry of Fisheries and the Environment, 1978.

Cannadine, David. *Ornamentalism: How the British Saw Their Empire.* London: Allen Lane, 2001.

Carlson, Andrew R. *German Foreign Policy, 1890–1914, and Colonial Policy to 1914: A Handbook and Annotated Bibliography.* Metuchen, NJ: Scarecrow Press, 1970.

Carruthers, Jane. "Game Protection in the Transvaal 1846–1926." PhD diss., University of Cape Town, 1988.

————. "Creating a National Park, 1910 to 1926." *Journal of Southern African Studies* 15, no. 2 (January 1989): 188–216.

————. *The Kruger National Park: A Social and Political History.* Pietermaritzburg, South Africa: University of Natal Press, 1995.

Cartmill, Matt. *A View to a Death in the Morning: Hunting and Nature through History.* Cambridge, MA: Harvard University Press, 1993.

Ceballos, Gerardo, and Laura Márquez Valdelamar, eds. *Las aves de México en peligro de extinción.* Mexico City: Instituto de Ecología Universidad Nacional Autónoma de México, 2000.

Chenevix Trench, Charles. *The Poacher and the Squire: A History of Poaching and Game Preservation in England.* London: Longmans, 1967.

Cherfas, Jeremy. *The Hunting of the Whale.* London: Penguin, 1989.

Coffey, David J. *Dolphins, Whales and Porpoises: An Encyclopedia of Sea Mammals.* New York: Collier Books, 1977.

Colinvaux, Paul A. *Why Big Fierce Animals Are Rare: An Ecologist's Perspective.* Princeton, NJ: Princeton University Press, 1979.

Committee for Whaling Statistics. *International Whaling Statistics.* Vols. 1–96. Oslo: Committee for Whaling Statistics, 1930–1988.

Connery, Robert H. *Governmental Problems in Wild Life Conservation.* New York: Columbia University Press, 1935.

Connor, Richard C., and Dawn Micklethwaite Peterson. *The Lives of Whales and Dolphins.* New York: Henry Holt, 1994.

Conrad, Jon M. "Bioeconomics and the Bowhead Whale." *Journal of Political Economy* 97, no. 4 (1989): 974–87.

"Conservación y preservación de regiones naturales y lugares históricos (Mexico)." *Diario de la VIII conferencia internacional americana,* no. 2 (December 7, 1938): 142–56.

Conservation International. *Mexico's Living Endowment: An Overview.* 1989.

Conservation of Nature and Natural Resources in Modern African States: Report of a Symposium Organized by CCTA and IUCN and Held under the Auspices of FAO and UNESCO at Arusha, Tanganyika, September 1961. Comp. Gerald Watterson. Morges, Switzerland: IUCN, 1963.

Council on Environmental Quality. *Wildlife in America: Contributions to an Understanding of American Wildlife and Its Conservation.* Washington, DC: U.S. Government Printing Office, 1978.

————. *Our Nation's Wetlands.* Washington, DC: U.S. Government Printing Office, 1979.

Cranworth, Lord. *Profit and Sport in British East Africa: Being a Second Edition, Revised and Enlarged, of "A Colony in the Making."* London: Macmillan, 1919.

Curry-Lindahl, K. *Let Them Live: A World Wide Survey of Animals Threatened with Extinction.* New York: Morrow, 1972.

Curtin, Philip D. *The Image of Africa: British Ideas and Action, 1780–1850.* Madison: University of Wisconsin Press, 1964.

Curtin, Philip D., Steven Feierman, Leonard Thompson, and Jan Vansina. *African History: From Earliest Times to Independence.* London: Longman, 1995.

Dahl, T. E. *Wetlands Losses in the United States 1780's to 1980's.* Washington, DC: U.S. Department of the Interior, Fish and Wildlife Services, 1990.

Dalleo, Peter T. "The Somali Role in Organized Poaching in Northeastern Kenya, c. 1909–1939." *International Journal of African Historical Studies* 12, no. 3 (1979): 472–82.

Darling, Jay N. *Ding's Half Century.* Edited by John M. Henry. New York: Duell, Sloan and Pearce, 1962.

Davis, Lance E., Robert E. Gallman, and Karin Gleiter. *In Pursuit of Leviathan: Technology, Institutions, Productivity, and Profits in American Whaling, 1816–1906.* Chicago: University of Chicago Press, 1997.

Davis, Lance E., Robert E. Gallman, and Teresa D. Hutchins. "The Decline of U.S. Whaling: Was the Stock of Whales Running Out?" *Business History Review* 62, no. 4 (1988): 569–95.

Day, Albert M. *North American Waterfowl.* New York: Stackpole and Heck, 1949.

Day, David. *The Whale War.* London: Routledge and Kegan Paul, 1987.

Dehler, Gregory J. "An American Crusader: William Temple Hornaday and Wildlife Protection in America, 1840–1940." PhD diss., Lehigh University, 2001.

Delany, M. J., and D. C. D. Happold. *Ecology of African Mammals.* London: Longman, 1979.

Dorsey, Kurkpatrick. *The Dawn of Conservation Diplomacy: U.S.-Canadian Wildlife Protection Treaties in the Progressive Era.* Seattle: University of Washington Press, 1998.

Dorst, Jean. *The Migrations of Birds.* Boston: Houghton Mifflin, 1962.

Doughty, Robin W. *Feather Fashions and Bird Preservation.* Berkeley: University of California Press, 1975.

Dovers, Stephen, Ruth Edgecombe, and Bill Guest, eds. *South Africa's Environmental History: Cases & Comparisons.* Athens: Ohio University Press, 2002.

Drummond, Henry. *Tropical Africa.* London: Hodder and Stoughton, 1889.

Dunlap, Thomas R. *Saving America's Wildlife.* Princeton, NJ: Princeton University Press, 1988.

Dunn, Erica H., Michael D. Cadman, and J. Bruce Falls, eds. *Monitoring Bird Populations: The Canadian Experience.* Occasional papers no. 95. Ottawa: Canadian Wildlife Services, 1997.

Eliot, Sir Charles. *The East Africa Protectorate.* 1905. London: Frank Cass, 1966.

Elliot, G. H. "The Failure of the IWC, 1946–1966." *Marine Policy* 3, no. 2 (April 1979): 149–55.

Ellis, Richard. *The Book of Whales.* New York: Alfred A. Knopf, 1980.

———. *Men and Whales.* New York: Alfred A. Knopf, 1991.

Elphick, Jonathan, gen. ed. *Collins Atlas of Bird Migration: Tracing the Great Journeys of the World's Birds.* London: HarperCollins, 1995.

Escalante Pliego, Patricia, Adolfo G. Navarro Sigüenza, and A. Townsend Peterson. "A Geographic, Ecological, and Historical Analysis of Land Bird Diversity in Mexico." In *Biological Diversity of Mexico: Origins and Distribution,* edited by T. P. Ramamoorthy, Robert Bye, Antonio Lot, and John Fa, 281–307. New York: Oxford, 1993.

Estes, Richard Despard. *The Behavior Guide to African Mammals, Including Hoofed Mammals, Carnivores, Primates.* Berkeley: University of California Press, 1991.

Evans, Peter G. H. *The Natural History of Whales and Dolphins.* New York: Facts on File Publications, 1987.

Fitter, Richard. *Vanishing Wild Animals of the World.* New York: Franklin Watts, 1968.

———. *The Penitent Butchers.* London: Collins, 1978.

Flader, Susan, and J. Baird Callicott, eds. *The River of the Mother of God and Other Essays.* Madison: University of Wisconsin Press, 1991.

Flinterman, Cees, Barbara Kwiatkowska, and Johan G. Lammers, eds. *Transboundary Air Pollution: International Legal Aspects of the Co-operation of States.* Dordrecht, the Netherlands: Martinus Nijhoff, 1986.

Forbush, Edward. *Useful Birds and Their Protection.* Boston: Wright and Potter, 1908.

Ford, John. *The Role of Trypanosomiasis in African Ecology: A Study of the Tsetse Fly Problem.* Oxford: Clarendon, 1971.

———. "African Trypanosomiasis: An Assessment of the Tsetse-Fly Problem Today." In *African Environment: Problems and Perspectives,* edited by Paul Richards, 67–72. London: International African Institute, 1975.

Foster, Janet. *Working for Wildlife: The Beginning of Preservation in Canada.* Toronto, Canada: University of Toronto Press, 1998.

Francis, Daniel. *A History of World Whaling.* Harmondsworth, UK: Penguin, 1990.

Freeman, Milton M. R., and Urs. P. Kreuter, eds. *Elephants and Whales: Resources for Whom?* Basel, Switzerland: Gordon and Breach, 1994.

Friedheim, Robert L., ed. *Toward a Sustainable Whaling Regime.* Seattle: University of Washington Press, 2001.

Gabrielson, Ira N. *Wildlife Refuges.* New York: Macmillan, 1943.

Galicia, Daniel F. "Mexico's National Parks." *Ecology* 22 (January 1941): 107–10.

Giblin, James L. *The Politics of Environmental Control in Northeastern Tanzania, 1840–1940.* Philadelphia: University of Pennsylvania Press, 1992.

———. "Trypanosomiasis Control in African History: An Evaded Issue?" *Journal of African History* 31, no. 1 (1990): 59–80.

Gilbert, Erik. *Dhows and the Colonial Economy of Zanzibar, 1860–1970.* Athens: Ohio University Press, 2004.

Gissibl, Bernhard. "German Colonialism and the Beginnings of International Wildlife Preservation in Africa," *GHI-Bulletin,* Supplement 3 (2006): 121–43.

Glooschenko, W. A., C. Tarnocai, S. Zoltai, and V. Glooschenko. "Wetlands of Canada and Greenland." In *Wetlands of the World: Inventory, Ecology and Management,* edited by Dennis Whigham, Dagmar Dykyjová, and Slavomil Hejný, 415–514. Dordrecht, the Netherlands: Kluwer, 1993.

Goble, Dale D., and Eric T. Freyfogle. *Wildlife Law: Cases and Materials.* New York: Foundation Press, 2002.

González, Ambrosio, and Víctor Manuel Sánchez. *Los parques nacionales de México.* Mexico City: Ediciónes del Instituto Mexicano de Recursos Naturales Renovables, 1961.

Gottesman, Dan. "Native Hunting and the Migratory Birds Convention Act: Historical, Political and Ideological Perspectives." *Journal of Canadian Studies* 18, no. 3 (1983): 67–89.

Graham, Alistair. *The Gardeners of Eden.* London: George Allen and Unwin, 1973.

Graham, Frank, Jr. *Man's Dominion: The Story of Conservation in America.* Philadelphia: M. Evans, 1971.

Greenwalt, Lynn A. "The National Wildlife Refuge System." In *Wildlife and America: Contributions to an Understanding of American Wildlife and Its Conservation,* edited by Howard P. Brokaw, 399–412. Washington, DC: U.S. Government Printing Office, 1978.

Grieves, Forest L. "Leviathan, the International Whaling Commission and Conservation as Environmental Aspects of International Law." *Western Political Quarterly* 25 (1972): 711–25.

Griffith, Tom, and Libby Robin, eds. *Ecology and Empire: Environmental History of Settler Societies.* Edinburgh: Keele University Press, 1997.

Grove, Richard. "Early Themes in African Conservation: The Cape in the Nineteenth Century." In *Conservation in Africa: People, Policies and Practice,* edited by David Anderson and Richard Grove. Cambridge: Cambridge University Press, 1987: 21–39.

Gruvel, A. "La pêche aux grands cétacés sur la côte occidentale d'Afrique." *Comptes rendues hebdomadaires des séances de l'Académie des Sciences* 156 (1913): 1705.

————. "La chasse aux cétacés dans le monde. Son avenir dans les colonies françaises." *Revue scientifique: Revue rose* 63, no. 3 (February 1925): 65–71.

Grzimek, Bernard, and Michael Grzimek. *Serengeti Shall Not Die.* Trans. E. L. Rewald and D. Rewald. New York: E. P. Dutton, 1961.

Haskell, William. *The American Game Protective and Propagation Association: A History.* New York, 1937.

Hayden, Sherman Strong. *The International Protection of Wild Life.* New York: Columbia University Press, 1942.

Hays, Samuel P. *Conservation and the Gospel of Efficiency: The Progressive Conservation Movement, 1890–1920.* Cambridge, MA: Harvard University Press, 1959.

Hewitt, C. Gordon. "Conservation of Birds and Mammals in Canada." In *Conservation of Fish, Birds and Game: Proceedings at a Meeting of the Committee, November 1 and 2, 1915,* 141–51. Toronto, Canada: Methodist Book and Publishing House, 1916.

Hingston, R. W. G. "Proposed National Parks for Africa: A Paper Read at the Evening Meeting of the Society on 9 March 1931." *Geographical Journal* 77, no. 5 (May 1931): 402–6.

Hobley, C. W. "The London Convention of 1900." *Journal of the Society for the Preservation of the Fauna of the Empire,* n.s., part 20 (August 1933): 33–49.

Hochschild, Adam. *King Leopold's Ghost: A Story of Greed, Terror, and Heroism in Colonial Africa.* Boston: Houghton Mifflin, 1998.

Hoppe, Kirk Arden. *Lords of the Fly: Sleeping Sickness Control in British East Africa, 1900–1960.* Westport, CT: Praeger, 2003.

Hornaday, William T. *Our Vanishing Wild Life: Its Extermination and Preservation.* New York: New York Zoological Society, 1913.

————. *Thirty Years War for Wild Life: Gains and Losses in the Thankless Task.* Stamford, CT: Permanent Wild Life Protection Fund, 1931.

Huxley, Julian. *African View.* New York: Harper and Brothers, 1931.

————. *The Conservation of Wild Life and Natural Habitats in Central and East Africa: Report on a Mission Accomplished for UNESCO, July–September 1960.* Paris: UNESCO, 1961.

Inskipp, Tim, and Sue Wells. *International Trade in Wildlife.* London: Earthscan, 1979.

Jackson, Gordon. *The British Whaling Trade.* London: Adam and Charles Black, 1978.

Jacobs, Nancy. *Environment, Power, and Injustice: A South African History.* Cambridge: Cambridge University Press, 2003.

Jacoby, Karl. *Crimes against Nature: Squatters, Poachers, Thieves, and the Hidden History of American Conservation.* Berkeley: University of California Press, 2001.

Jenkins, J. T. *A History of the Whale Fisheries from the Basque Fisheries of the Tenth Century to the Hunting of the Finner Whale at the Present Date.* Port Washington, NY: Kennikat Press, 1921.

Jonsgård, Åge. *Biology of the North Atlantic Fin Whale Balaeoptera Physalus (L): Taxonomy, Distribution, Migration and Food.* Oslo: Universitetsforlaget, 1966.

Kelly, Nora. "In Wildest Africa: The Preservation of Game in Kenya 1895–1933." PhD diss., Simon Fraser University, 1978.

Kjekshus, Helge. *Ecology Control and Economic Development in East African History: The Case of Tanganyika, 1850–1950.* London: Heinemann, 1977.

Klemm, Cyril de. "Migratory Species in International Law." *Natural Resources Journal* 29 (Fall 1989): 935–78.

Knox, George A. *The Biology of the Southern Ocean.* Cambridge: Cambridge University Press, 1994.

Koponen, Juhani. *People and Production in Late Precolonial Tanzania: History and Structures.* Uppsala, Sweden: Scandinavian Institute of African Studies, 1988.

———. "War, Famine, and Pestilence in Late Precolonial Tanzania: A Case for a Heightened Mortality." *International Journal of African Historical Studies* 21, no. 4 (1988): 637–76.

Kreuter, Urs. P., and Randy T. Simmons. "Who Owns the Elephants? The Political Economy of Saving the African Elephant." In *Wildlife in the Marketplace: The Political Economy Forum,* edited by Terry Anderson and Peter Hill, 147–65. Lanham, MD: Rowman and Littlefield, 1995.

Kunz, George Frederick. *Ivory and the Elephant in Art, in Archaeology, and in Science.* Garden City, NY: Doubleday, 1916.

Laws, R. M., I. S. C. Parker, and R. C. B. Johnstone. *Elephants and Their Habitats: The Ecology of Elephants in North Bunyoro, Uganda.* Oxford: Clarendon Press, 1975.

Lawyer, George A. *Federal Protection of Migratory Birds.* Yearbook of the Department of Agriculture, no. 785. Washington, DC: U.S. Government Printing Office, 1919.

Leach, Melissa, and Robin Mearns, eds. *The Lie of the Land: Challenging the Received Wisdom on the African Environment.* Portsmouth, NH: Heinemann, 1996.

Leopold, Aldo. *Game Management.* New York: Charles Scribner's Sons, 1933.

———. *Round River: From the Journals of Aldo Leopold.* Edited by Luna B. Leopold. London: Oxford University Press, 1953.

Leopold, Starker. *Wildlife of Mexico: The Game Birds and Mammals.* Berkeley: University of California Press, 1959.

Lewis, Keri. "Negotiating Nature: The Convention on Nature Protection and Wildlife Preservation in the Western Hemisphere, 1930–1950." PhD diss., University of New Hampshire, 2007.

Lewis, L. A., and L. Berry. *African Environments and Resources.* Boston: Unwin Hyman, 1988.

Lincoln, Frederick. *The Migration of American Birds.* New York: Doubleday, 1939.

Loo, Tina. *States of Nature: Conserving Canada's Wildlife in the Twentieth Century.* Vancouver, Canada: University of British Columbia, 2006.

López Portillo y Ramos, Manuel, comp. *El medio ambiente en México: Temas, problemas, alternativas.* Mexico City: Fondo de Cultura Económica, 1982.

Luard, Nicholas. *The Wildlife Parks of Africa.* London: Michael Joseph, 1985.

Lyster, Simon. *International Wildlife Law: An Analysis of International Treaties Concerned with the Conservation of Wildlife.* Cambridge: Grotius Publications, 1985.

———. "The Convention on the Conservation of Migratory Species of Wild Animals (The 'Bonn Convention')." *Natural Resources Journal* 29 (Fall 1989): 979–1000.

Macias, Luis. "Wildlife Problems in Mexico." In *Transactions of the Fourteenth North American Wildlife Conference,* 9–16. Washington, DC: Wildlife Management Institute, 1949.

MacKenzie, John M. *The Empire of Nature.* Manchester, UK: Manchester University Press, 1988.

———. *Imperialism and the Natural World.* Manchester, UK: Manchester University Press, 1990.

Maddox, Gregory, James Giblin, and Isaria Kimambo, eds. *Custodians of the Land: Ecology & Culture in the History of Tanzania.* Athens: Ohio University Press, 1996.

Mann, Janet, Richard C. Connor, Peter L. Tyack, and Hal Whitehead, eds. *Cetacean Societies: Field Studies of Dolphins and Whales.* Chicago: University of Chicago Press, 2000.

Marks, Stuart. *Imperial Lion: Human Dimensions of Wildlife Management in Central Africa.* Boulder, CO: Westview Press, 1984.

Matthews, Leonard Harrison. *The Whale.* New York: Simon and Schuster, 1968.

———. *The Natural History of the Whale.* New York: Columbia University Press, 1978.

Mawer, Granville Allen. *Ahab's Trade: The Saga of South Seas Whaling.* New York: St. Martin's Press, 1999.

McCaddin, Joe, ed. *Duck Stamps and Prints: The Complete Federal and State Editions.* New York: Hugh Lauter Levin Associates, 1988.

McCann, James. *Green Land, Brown Land, Black Land.* Portsmouth, NH: Heinemann, 1999.

McCormick, John. *Reclaiming Paradise: The Global Environmental Movement.* Bloomington: Indiana University Press, 1989.

McHugh, L. "Rise and Fall of World Whaling: The Tragedy of the Commons Illustrated." *Journal of International Affairs* 31, no. 1 (1977): 23–26.

McVay, Scott. "The Last of the Great Whales." *Scientific American* 215, no. 2 (August 1966): 3–11.

Meine, Curt. *Aldo Leopold: His Life and Work.* Madison: University of Wisconsin Press, 1988.

Melville, Herman. *Moby-Dick; or, The Whale.* 1851. Indianapolis, IN: Bobbs-Merrill, 1964.

Mighetto, Lisa. *Wild Animals and American Environmental Ethics.* Tucson: University of Arizona Press, 1991.

Migratory Bird Conservation Commission. *2000 Annual Report: Migratory Bird Conservation Commission.* Washington, DC: Migratory Bird Conservation Commission, 2000.

The Migratory Birds Convention Act and Federal Regulations for the Protection of Migratory Birds. Ottawa: F. A. Acland, 1927.

The Migratory Birds Convention Act and Federal Regulations for the Protection of Migratory Birds (Revised Statutes of Canada). Ottawa: F. A. Acland, 1933.

Miner, Jack. *Jack Miner and the Birds, and Some Things I Know about Nature.* Chicago: Reilly and Lee, 1923.

Molitor, Michael R., ed. *International Environmental Law: Primary Materials.* Deventer, the Netherlands: Kluwer, 1991.

Moore, E. D. *Ivory: Scourge of Africa.* New York: Harper and Brothers, 1931.

Moser, Michael E. "Priorities for the Conservation of Migratory Waterfowl." In *Conserving Migratory Birds,* edited by T. Salathé. ICBP Technical Publication no. 12, 361–74. Cambridge: International Council for Bird Preservation, 1991.

Mutwira, Roben. "Southern Rhodesian Wildlife Policy (1890–1953): A Question of Condoning Game Slaughter?" *Journal of Southern African Studies* 15, no. 2 (January 1989): 250–62.

Nash, Roderick Frazier. *Wilderness and the American Mind.* 1967. New Haven, CT: Yale University Press, 2001.

Navid, Daniel. "The International Law of Migratory Species: The Ramsar Convention." *Natural Resources Journal* 29 (Fall 1989): 1001–16.

Nwulia, Moses D. E. *Britain and Slavery in East Africa*. Washington, DC: Three Continents Press, 1975.

O'Brien, Sharon. "Undercurrents in International Law: A Tale of Two Treaties." *Canada–United States Law Journal* 9, no. 1 (1985): 1–57.

Ofcansky, Thomas P. "A History of Game Preservation in British East Africa, 1895–1963." PhD diss., West Virginia University, 1981.

———. *Paradise Lost: A History of Game Preservation in East Africa*. Morgantown: West Virginia University Press, 2002.

Oliver, Roland, and Michael Crowder, eds. *Cambridge Encyclopedia of Africa*. Cambridge: Cambridge University Press, 1981.

Olmsted, Ingrid. "Wetlands of Mexico." In *Wetlands of the World: Inventory, Ecology and Management*, edited by Dennis Whigham, Dagmar Dykyjová, and Slavomil Hejný, 637–75. Dordrecht, the Netherlands: Kluwer, 1993.

Orr, Oliver H., Jr. *Saving American Birds: T. Gilbert Pearson and the Founding of the Audubon Movement*. Gainesville: University Press of Florida, 1992.

Ossa, Helen. *They Saved Our Birds: The Battle Won and the War to Win*. New York: Hippocrene Books, 1973.

Pakenham, Thomas. *The Scramble for Africa: White Man's Conquest of the Dark Continent from 1876 to 1912*. New York: Avon Books, 1991.

Palmer, T. S., W. F. Bancroft, and Frank L. Earnshaw. "Game Laws of 1913: A Summary of the Provisions Relating to Seasons, Export, Sale, Limits, and Licenses." *Bulletin of the U.S. Department of Agriculture*, no. 22 (September 16, 1913): 1–51.

Parker, I. S. C. "The Tsavo Story: An Ecological Case History." In *Management of Large Mammals in African Conservation Areas: Proceedings of a Symposium Held in Pretoria, South Africa, 29–30 April 1982*, edited by R. Norman Owen-Smith, 37–49. Pretoria, South Africa: HAUM, 1983.

Parker, I. S. C., and Mohamed Amin. *Ivory Crisis*. London: Chatto and Windus, 1983.

Payne, Roger. *Among Whales*. New York: Scribner, 1995.

Pearson, T. Gilbert. "What Is the Audubon Society?" *National Association of Audubon Societies*, circular no. 16 (1930).

Perras, Anna. *Carl Peters and German Imperialism, 1856–1918: A Political Biography*. Oxford: Clarendon Press, 2004.

Peterson, Dale. *Eating Apes*. Berkeley: University of California Press, 2003.

Phillips, John C. *Migratory Bird Protection in North America: The History of Control by the United States Federal Government and a Sketch of the Treaty with Great Britain*. Cambridge, MA: American Committee for International Wild Life Protection, 1934.

Pollock, Norman Charles. *Animals, Environment, and Man in Africa*. Farnborough, UK: Saxon House, 1974.

Price, Jennifer. *Flight Maps: Adventures with Nature in Modern America*. New York: Basic Books, 1999.

Pringle, John A. *The Conservationists and the Killers: The Story of Game Protection and the Wildlife Society of Southern Africa*. Cape Town: T. V. Bulpin, 1982.

Ranger, Terence. *Voices from the Rocks: Nature, Culture, and History in the Matopos Hills of Zimbabwe*. Oxford: James Currey, 1999.

Redington, Paul. *Report of the Chief of the Bureau of Biological Survey, U.S. Department of Agriculture*. Washington, DC: U.S. Government Printing Office, 1931.

Reiger, John F. *American Sportsmen and the Origins of Conservation*. Corvallis: Oregon State University Press, 2001.

"Relación de los parques nacionales que han sido declarados desde la creación del Departamento Forestal y de Caza y Pesca hasta el 24 de noviembre de 1939." *México Forestal* 17 (July-December 1939): 67–74.

Ritvo, Harriet. *The Animal Estate: The English and Other Creatures in the Victorian Age*. Cambridge, MA: Harvard University Press, 1987.

———. "Destroyers and Preservers: Big Game in the Victorian Empire." *History Today* 52, no. 1 (January 2002): 33–39.

Robertson, R. B. *Of Whales and Men*. New York: Alfred A. Knopf, 1961.

Rose, Debra. "The Politics of Mexican Wildlife: Conservation, Development, and the International System." PhD diss., University of Florida, 1993.

Rowe, L. S., and Pedro de Alba, eds. *Documentary Material on Nature Protection and Wild Life Preservation in Latin America*. Washington, DC: Pan American Union, 1940.

———. *Preliminary Draft Convention on Nature Protection and Wild Life Preservation in the Western Hemisphere*. Washington, DC: Pan American Union, 1940.

Rüster, Bernd, and Bruno Simma, eds. *International Protection of the Environment: Treaties and Related Documents*. Vols. 1–18. Dobbs Ferry, NY: Oceana Publications, 1975–1983.

Salathé, T., ed. *Conserving Migratory Birds*. ICBP Technical Publication no. 12. Cambridge: International Council for Bird Preservation, 1991.

Sands, Philippe, ed. *Greening International Law*. New York: New Press, 1994.

Saunders, George B. "Waterfowl Wintering Grounds of Mexico." In *Transactions of the Seventeenth North American Wildlife Conference*, 89–100. Washington, DC: Wildlife Management Institute, 1952.

Schevill, William E., ed. *The Whale Problem: A Status Report*. Cambridge, MA: Harvard University Press, 1974.

Schilling, Carl G. *With Flashlight and Rifle*. London: Hutchinson, 1905.

Schmitt, Frederick P. *Mark Well the Whale! Long Island Ships to Distant Seas.* Port Washington, NY: Kennikat Press, 1971.

Schmitt, Frederick P., Cornelis de Jong, and Frank H. Winter. *Thomas Welcome Roys: America's Pioneer of Modern Whaling.* Charlottesville: University Press of Virginia, 1980.

Sheldon, H. P. "Game Laws for the Season 1935–36: A Summary of Federal, State, and Provincial Statutes." *Bulletin of the U.S. Department of Agriculture: Farmers' Bulletin,* no. 1755 (1936).

Sheriff, Abdul. *Slaves, Spices, and Ivory in Zanzibar.* Athens: Ohio University Press, 1987.

Shillington, Kevin. *History of Africa.* New York: St. Martin's Press, 1995.

Simonian, Lane. *Defending the Land of the Jaguar: A History of Conservation in Mexico.* Austin: University of Texas Press, 1995.

Small, George L. *The Blue Whale.* New York: Columbia University Press, 1971.

Sosa, Antonio H. "Los parques nacionales de México." *México Forestal* 42 (November-December 1968): 17–30.

Starbuck, Alexander. *The History of the American Whale Fishery.* 1877. Secaucus, NJ: Castle Books, 1989.

Steinhart, Edward I. "The Imperial Hunt in Colonial Kenya, c. 1880–1909." In *Animals in Human Histories: The Mirror of Nature and Culture,* edited by Mary J. Henninger-Voss, 144–81. Rochester, NY: University of Rochester Press, 2002.

———. *Black Poachers, White Hunters: A Social History of Hunting in Colonial Kenya.* Athens: Ohio University Press, 2006.

Stevenson-Hamilton, J. *South African Eden: From Sabi Game Reserve to Kruger National Park.* London: Cassell, 1937.

Stoett, Peter J. *The International Politics of Whaling.* Vancouver, Canada: University of British Columbia Press, 1997.

Stone, Jeffrey, ed. *The Exploitation of Animals in Africa: Proceedings of a Colloquium at the University of Aberdeen, March 1987.* Aberdeen, Scotland: Aberdeen University African Studies Group, 1988.

Strong, Richard, Joseph C. Bequaert, and L. R. Cleveland. *Report on the Available Evidence Showing the Relation of Game to the Spread of Tsetse Fly Borne Diseases in Africa.* Cambridge, MA: American Committee for International Wild Life Protection, 1931.

Tacha, Thomas C., and Clait E. Braun, *Migratory Shore and Upland Game Bird Management in North America.* Washington, DC: U.S. Fish and Wildlife Service, 1994.

Taylor, A. J. P. *Germany's First Bid for Colonies, 1884–1885: A Move in Bismarck's Foreign Policy.* London: Macmillan, 1938.

Thompson, E. P. *Whigs and Hunters: The Origin of the Black Act.* London: Allen Lane, 1975.

Tillman, Michael F., and Gregory P. Donovan, eds. *Historical Whaling Records: Including the Proceedings of the International Workshop on Historical Whaling Records, Sharon, Massachusetts, September 12–16, 1977.* Cambridge, MA: International Whaling Commission, 1983.

Tinker, Ben. *Mexican Wilderness and Wildlife.* Austin: University of Texas Press, 1978.

Titley, E. Brian. *A Narrow Vision: Duncan Campbell Scott and the Administration of Indian Affairs in Canada.* Vancouver, Canada: University of British Columbia, 1986.

Tjader, Richard. *The Big Game of Africa.* New York: D. Appleton, 1910.

Tønnessen, J. N., and A. O. Johnsen. *The History of Modern Whaling.* Berkeley: University of California Press, 1982.

Trefethen, James B. *An American Crusade for Wildlife.* New York: Winchester Press, 1975.

United Kingdom. Parliament. *Correspondence Relating to the Preservation of Wild Animals in Africa.* Cd. 3189. November 1906.

———. *Further Correspondence Relating to the Preservation of Wild Animals in Africa.* Cd. 4472. January 1909.

———. *Further Correspondence Relating to the Preservation of Wild Animals in Africa.* Cd. 5136. June 1910.

———. *Further Correspondence Relating to the Preservation of Wild Animals in Africa.* Cd. 5775. July 1911.

———. *Further Correspondence Relating to the Preservation of Wild Animals in Africa.* Part 2/7822. 1902.

———. *Further Correspondence Respecting the Preservation of Wild Game in Africa.* Part 3/8384. 1904.

———. *Further Correspondence Respecting the Preservation of Wild Game in Africa.* Part 4/8991. 1905.

———. *Further Correspondence Respecting East Africa.* 1897–1900.

U.S. Congress. House. Committee on Agriculture. *Migratory Bird Refuges and Public Shooting Grounds: Hearings before the Committee on Agriculture.* 67th Cong., 2nd sess., Series T. February 16 and 17, 1922.

———. *Amend the Migratory Bird Treaty Act with Respect to Bag Limits: Hearings before the Committee on Agriculture.* 71st Cong., 2nd sess., January 27, 28, and 29, 1930.

U.S. Congress. Senate. Committee on Agriculture and Forestry. *Agriculture Appropriation Bill, 1917 (Protection of Migratory Birds): Hearing before the Committee on Agriculture and Forestry.* 64th Cong., 1st sess., May 22, 1916.

————. *Bear River Migratory Bird Refuge, a Bill to Establish the Bear River Migratory Bird Refuge and a Bill Authorizing the Establishment of a Migratory Bird Refuge at Bear River Bay, Great Salt Lake, Utah: Hearings before the Committee on Agriculture and Forestry.* 70th Cong., 1st. sess., February 14, 1928.

————. *Migratory Bird Conservation Act, a Bill to More Effectively Meet the Obligations of the United States under the Migratory Bird Treaty with Great Britain by Lessening the Dangers Threatening Migratory Game Birds from Drainage and Other Causes, by the Acquisition of Areas of Land and of Water to Furnish in Perpetuity Reservations for the Adequate Protection of Such Birds, and by Providing Funds for the Establishment of Such Areas, Their Maintenance and Improvement, and for Other Purposes: Hearings before the Committee of Agriculture and Forestry.* 70th Cong., 1st sess., 1928.

U.S. Congress. Senate. Subcommittee of the Committee on Agriculture and Forestry. *Resolution Requesting All Files and Data in Department of Agriculture Relating to Convention for Protection of Migratory Birds: Hearings before a Subcommittee of the Committee on Agriculture and Forestry.* 69th Cong., 2nd sess., January 29 and February 8, 1927.

————. *Protocol Amending the 1916 Convention for the Protection of Migratory Birds.* 104th Cong., 2d sess., 1996. S. Treaty Doc. 28.

————. *Protocol with Mexico Amending Convention for Protection of Migratory Birds and Game Mammals.* 105th Cong., 1st sess., 1997. S. Treaty Doc. 26.

U.S. Department of Agriculture. *National Reservations for the Protection of Wild Life.* Department circular 87. Washington, DC: U.S. Department of Agriculture, 1912.

————. *Service and Regulatory Announcements (August 21, 1916).* Bureau of Biological Survey S.R.A.-B.S. 11. Washington, DC: U.S. Government Printing Office, 1916.

————. *The Migratory Bird Treaty: Decision of the Supreme Court of the United States Sustaining the Constitutionality of the Migratory Bird Treaty and Act of Congress to Carry It into Effect.* Department circular 102. Washington, DC: U.S. Department of Agriculture, 1920.

————. *The Migratory Bird Treaty Act: "United States vs. Joseph H. Lumpkin."* Department circular 202. Washington, DC: U.S. Department of Agriculture, 1922.

Vail, Leroy. "Ecology and History: The Example of Eastern Zambia." *Journal of Southern African Studies* 3, no. 2 (1977): 129–55.

Valk, A. G. van der. "The Prairie Potholes of North America." In *The World's Largest Wetlands: Ecology and Conservation,* edited by Lauchlan Fraser and Paul A. Keddy, 393–423. Cambridge: Cambridge University Press, 2005.

Vargas Márquez, Fernando. *Parques nacionales de México y reservas equivalentes: Pasado, presente y futuro.* Mexico City: Instituto de Investigaciónes Económicas– UNAM, 1984.

Vencil, Betsy. "The Migratory Bird Treaty Act—Protecting Wildlife on Our National Refuges—California's Kesterson Reservoir, a Case in Point." *Natural Resources Journal* 26, no. 3 (Summer 1986): 609–27.

Villiers, A. J. *Whaling in the Frozen South: Being the Story of the 1923–24 Norwegian Whaling to the Antarctic.* 1925. New York: Bobbs-Merrill, 1931.

Waller, Richard. "Emutai: Crisis and Response in Maasailand 1883–1902." In *The Ecology of Survival: Case Studies from Northeast African History,* edited by Douglas H. Johnson and David M. Anderson, 73–112. London: Lester Crook, 1988.

Warren, Louis S. *The Hunter's Game: Poachers and Conservationists in Twentieth-Century America.* New Haven, CT: Yale University Press, 1997.

Wilson, Derek, and Peter Ayerst. *White Gold: The Story of African Ivory.* London: Heinemann, 1976.

Wilson, Robert M. "Directing the Flow: Migratory Waterfowl, Scale, and Mobility in Western North America." *Environmental History* 7, no. 2 (April 2002): 247–66.

Zinzer, Juan. "The Mexican Wildlife Situation." In *Proceedings of the First North American Wildlife Conference.* Washington, DC: North American Wildlife Conference, 1938.

Ziswiler, Vinzenz. *Extinct and Vanishing Animals: A Biology of Extinction and Survival.* New York: Springer-Verlag, 1967.

Index

www.ingramcontent.com/pod-product-compliance
Lightning Source LLC
Chambersburg PA
CBHW021855020426
42334CB00013B/347